Dynamics of Magnetically Trapped Particles

For further volumes:
http://www.springer.com/series/5664

Astrophysics and Space Science Library

Juan G. Roederer • Hui Zhang

Dynamics of Magnetically Trapped Particles

Foundations of the Physics of Radiation Belts and Space Plasmas

Second Edition

Springer

Juan G. Roederer
Hui Zhang
Geophysical Institute
University of Alaska-Fairbanks
Fairbanks, Alaska
USA

ISSN 0067-0057
ISBN 978-3-642-41529-6 ISBN 978-3-642-41530-2 (eBook)
DOI 10.1007/978-3-642-41530-2
Springer Heidelberg New York Dordrecht London

Library of Congress Control Number: 2013956932

Cover image: This diagram (rendered by Jerry Goldstein, Southwest Research Institute) provides a cross-sectional view of Earth's magnetosphere indicating its outermost interactions with the solar wind (purple boundary to the right) and showing the interior radiation belt regions. The inner part of the diagram illustrates the unexpected triple radiation structures discovered by NASA's Radiation Belt Storm Probes mission during September 2012. The radiation belts are torus-shaped volumes in three dimensions with white denoting highest electron fluxes and blue denoting lowest fluxes. The temporary appearance of a third belt signals the operation of a local acceleration mechanism, possibly inside the denser plasmasphere (green glow). (Courtesy of Daniel N. Baker, University of Colorado-Boulder)

Printed on acid-free paper

Springer is part of Springer Science+Business Media (www.springer.com)

J. G. R. dedicates his contribution to the memory of his most influential teachers: G. A. Dufour (Colegio Nacional J. A. Roca, Buenos Aires); L. A. Santaló (Universidad de Buenos Aires); and W. Heisenberg (Max Planck Institut für Physik, Göttingen)

H. Z. expresses her gratitude to Q.-G. Zong (Peking University); T. A. Fritz (Boston University); and D. Sibeck (NASA Goddard Space Flight Center), for their guidance and support in her career

Foreword

On the evening of 31 January 1958, in the Great Hall of the US National Academy of Sciences, James A. Van Allen, Wernher von Braun, and William Pickering held aloft the engineering model of the *Explorer I* spacecraft that had just been successfully launched into orbit by the United States. The discoveries made by instruments within this small artificial terrestrial satellite and the succeeding several space missions showed that the Earth's magnetic field is capable of confining energetic particles in a highly effective manner. Subsequent research showed that the region surrounding the Earth is not just a passive reservoir of trapped radiation, but is—in fact—an example of a powerful, efficient astronomical particle accelerator. Having such a remarkable accelerator in our cosmic backyard, so to speak, has made the discovery of Van Allen and coworkers all the more important to the broader discipline of astrophysics.

Several generations of researchers have grown up, in a scientific sense, grappling with the issue of how charged particles are accelerated, transported, and ultimately lost in the terrestrial magnetic environs. The explorer's guide to this fascinating realm of magnetospheric mystery has been the classic textbook of J. G. Roederer entitled *Dynamics of Geomagnetically Trapped Radiation* published in 1970. This book showed in clear, concise ways how the Earth's magnetic field was configured and showed how and why energetic particles moved in the complex way that they were observed to do. Since the pioneering work of Van Allen and coworkers, space research has shown that essentially all magnetized planets have variants of the Earth's radiation regions surrounding them. A gas giant planet like Jupiter has unimaginably intense radiation belt particle populations. A small, airless, magnetized body like Mercury does not have stable radiation belts in a strict sense, but nonetheless energetic electrons and ions spiral, bounce, and drift in that planetary magnetic field as much as the physical system will permit. Truly, the processes played out in the terrestrial magnetospheric laboratory have relevance to every plasma system human machines have visited directly. It is not an unreasonable extrapolation to conclude that the physics that has been worked out for the geomagnetic domain will have great applicability to many of the thousands of extra-solar system planets now discovered around other stars.

Since the publication of Roederer's book in 1970, there has been an exponential increase in the exploitation of space for human societal purposes. The fact that the space environment greatly affects operational spacecraft and many other technologies has given rise to the concept of "Space Weather". The need to understand the near-Earth radiation environment even to the point of accurate predictive modeling has made the value of Roederer's book even more obvious. Now a new NASA mission, the Radiation Belt Storm Probes (renamed the "Van Allen Probes" mission) has recently come on the scene. This amazing mission has already been reworking the observational landscape and it will require even deeper physical understanding of energetic particle behavior.

In light of the renewed interest and public fascination with radiation belt science, it is entirely fitting that a new version of the classic Roederer textbook be published. In these pages, Prof. Roederer and his colleague Hui Zhang have completely modernized and reworked the original book. Those familiar with the original text will recognize the structure and approach to particle motion and trapping properties. But the reader will also find new treatments, deeper physical insights, and better ties to broader plasma physical issues. Thus, this revamped version of the classic text provides a refreshing and timely new contribution to a field of study that got its start over five decades ago and, yet, is as important to space science as it ever was.

University of Colorado
Boulder, Colorado, USA

Daniel N. Baker
Director, Laboratory for
Astrophysics and Space Physics

Preface

The discovery in 1958 of magnetically trapped energetic particles surrounding planet Earth [1, 2] marked the beginnings of a new scientific discipline—Space Physics. Early US and Soviet satellites provided a wealth of initial data on the radiation belts; internationally coordinated satellite missions in the 1970s led to the study of relevant source, acceleration and diffusion processes. In the 1980s and 1990s, with the Pioneer, Galileo and Cassini missions, space physics turned its interest to the outer planets; terrestrial radiation belt studies fell somewhat into a hiatus. The situation, however, is changing, with several satellite missions like the NASA Van Allen Probes, the Canadian Outer Radiation Belt Injection, Transport, Acceleration, and Loss Satellite (ORBITALS) and the Japanese Energization and Radiation in Geospace (ERG) mission, all designed to enhance our understanding of radiation belt dynamics. After only a few months in orbit, the Van Allen Probes duo already has provided unsuspected and tantalizing results on the dynamics of geomagnetically trapped particles [3, 4]. This whole development—we may call it rejuvenation of radiation belt physics—will also require revitalizing the theoretical studies of all physical processes involved.

This book represents a thorough revision, expansion and update of Juan G. Roederer's 1970 "classic" *Dynamics of Geomagnetically Trapped Radiation* [5] (Russian translation: *Dinamika Radiatsiy, Zakhvatchennoi Geomagnitnim Polyem*, Publishing House Mir 1971, Moscow). It is complemented and amplified with material from more recent lecture notes from both authors and their research experience in the study of terrestrial and Jovian magnetospheres.

Like its old predecessor, *this is not a cookbook* with recipes that can be followed blindly to obtain some concrete results. Emphasis is not on the "what" but on the "why" of things out there in near-earth space. Our goal is not to tell but to explain, leading the student and beginning scientist on a path of *understanding* the underlying physics—rather than just offering a collection of laws and rules and their mathematical expressions. Our hope is to provide the reader with the mental tools to find out, not only what physical variables are related to each other, but *why* they are related at all, and to seek new causal relationships as new experimental results become available. In short, old equations can be looked up in books—new ones must be created by the brain. Essential theoretical complements for the present volume are Northrop's classic booklet [6], mathematically rigorous but highly compressed (it has everything!), and the detailed and highly didactical treatise by Rossi and Olbert [7] (it's all relativistic!).

The senior author (JGR) thanks his wife Beatriz for her infinite patience when, after almost 20 years of "active retirement" and his defection into a biology-related research field, he decided to come back to the good old topic on which he once built his career as a young scientist. The junior author (HZ) wishes to express her gratitude to Robert McCoy, Director of the Geophysical Institute, for his encouragement and the financial support from the Geophysical Institute for her time to participate in this enterprise. And both authors are immensely grateful to Solène Lejosne, who during the course of her thesis work at the University of Toulouse (France) [8] had contacted JGR with a number of very pointed questions and comments on his old book. On that basis we took the fortunate decision to ask Ms. Lejosne to help us as an unbiased critical reviewer from a grad-student's perspective, which she did with painstaking precision and dedication. Her collaboration confirmed the long-suspected fact that young scientists are far more valuable judges of the pedagogical value of a written piece than older experts!

The prerequisites for understanding the physics discussed in this book are some basic understanding of the magnetosphere and its principal regions and perturbations, and a reasonably good knowledge of electromagnetism, up to and including Maxwell's equations. To remind the reader of some necessary tools from that latter discipline, we have written Appendix A.1 which, rather than a mere presentation of algebraic, differential and geometric vector relationships between magnetic and electric field quantities needed in the main text, presents a discussion of the *physical meaning and field-topological consequences* of such relationships. We briefly elaborate in a perhaps less traditional way on Maxwell's equations and some important conceptual aspects that are germane to a better understanding of cause-and-effect relationships in the physics of radiation belts and magnetospheric plasmas. We strongly recommend that Appendix A.1 be read first.

Chapter 1 develops the adiabatic theory of particle motion from first principles. It is presented as a prime example of *physics as the art of modeling*, in which a complex real system is replaced by a highly simplified *virtual*, i.e., imagined one. In our case, the original system is a charged particle in complex multi-periodic motion in a magnetic field, replaced by a so-called *guiding center particle* (a model!) of equal mass and charge, moving in a smooth, uncomplicated way (having averaged out the smaller scale turns and loops of the real particle). The instantaneous position of this virtual particle is the *guiding center*, a point in space whose coordinates depend on the local magnetic field and the properties of the real particle. In addition, we show why the guiding center particle is endowed with a *magnetic moment* which impersonates average electromagnetic properties of the rapidly cycling motion of the original particle. The notion of *drift velocity* is introduced, and other fundamental physical magnitudes germane to adiabatic motion are defined; drift velocities are classified into zero order (independent of the particle's dynamic properties), first order and higher order. The *first adiabatic invariant* is defined and a simple demonstration of its near-constancy under certain restrictions (the adiabatic conditions) is given. We then discuss in detail what happens to a charged particle in its cyclotron motion when the magnetic field is slowly time-dependent, demonstrating the fundamental role of the often neglected magnetic vector potential. Examples are given for zero and first order drifts in simple field configurations. However simple, these examples (e.g., ion pick-up, adiabatic breakdown, closed vs. open drift orbits and their separatrix, co-rotating vs. convecting regions in an externally imposed electric field) illustrate some important basic properties of particles trapped in the equatorial region of the magnetosphere. In the course of this chapter, we emphasize that *magnetic field lines* are purely geometric entities, useful for one's mental representation of magnetic field configurations but, like the notion of a guiding center particle, devoid of any physical reality; nonetheless we give a phenomenological definition of field line velocity—again putting to good use the vector potential and its local time variation.

Chapter 2 formalizes the definition of drift velocity in general terms for arbitrary field configurations; some specific second order drifts are discussed. We show *why* guiding center particles follow curved magnetic field lines (a non-trivial fact) and discuss the conditions for that to happen. Since drifts are, by definition, perpendicular to the local magnetic field vector, we then examine the *parallel motion* of the guiding center, introducing the all-important concepts of *particle trapping*, *mirror points* and periodic *bounce motion*. In discussing the concept of bounce period we show that, like a pendulum at rest, even non-bouncing equatorial (90°-pitch angle) particles have an intrinsic bounce period (with which external perturbations can resonate). In the second part of this chapter, we take a close look at the guiding center particle's parallel acceleration and total kinetic energy change along a magnetic field line in the presence of a field-aligned electric field, deriving the *energy equation* and the so-called *betatron and Fermi accelerations*. The chapter ends with a thorough discussion of the effects of given potential (conservative) *parallel electrostatic fields* on a particle's bounce motion, identifying distinct regions of behavior in a parallel/perpendicular velocity map—a subject of importance in auroral physics.

Chapter 3, based mainly on the 1970 edition [5], defines the fundamental concept of trapped particle *drift shells*. Two additional field-geometric concepts are introduced and added to the concept of guiding center: the *guiding field line* and the *guiding drift shell*. All this leads to the definition of the *second adiabatic invariant* and several related simplified expressions thereof valid for some special cases. The conservation of the second invariant is demonstrated (in Appendix A.2). Using this theorem, it is possible to trace drift shells in general fields, and despite the "cookbook" disclaimer at the beginning of this Preface, some practical recipes are given for numerical methods to accomplish this. The concepts of *shell splitting* and *pseudo-trapping* in azimuthally asymmetric fields are discussed, and consequences for particle trapping and diffusion are mentioned. Several examples are analyzed in detail and analytical expressions for near-equatorial particles are given. A special look is taken at the dipole field as a first approximation of the geomagnetic field. Old but still much used quantities are introduced and discussed, such as the *L-value* and the system of *invariant coordinates*. We examine the situation of near-equatorial particles (pitch angles near 90°), for whose drift orbits analytical relationships can be written down for first-order magnetospheric field approximations. The last part of this chapter deals with slowly time-varying fields and the resulting effects on drift shells. We introduce the *third adiabatic invariant*, a purely field-geometric quantity (the magnetic flux enclosed by a drift shell), and demonstrate (in Appendix A.3) its constancy under adiabatic conditions. The process during the adiabatic change is examined "under a microscope", showing that adiabatic constancy is not absolute but valid only when averaged over a drift period: under time-dependent conditions, identical particles on the same drift shell share a common drift shell only at times that are integer multiples of their drift period. Without proof, we mention that this really is also true for the other two adiabatic invariants and their related periodicities. A generalized L-parameter, or "*L-star*" is introduced, and a general method for its calculation is presented.

Chapter 4 comes down from the lofty heights of pure theory and introduces the concept of flux of an ensemble of adiabatically behaving particles. It defines the corresponding physical concepts of differential *directional flux* and that of the ubiquitous *distribution function* as average quantities linking actual particle ensembles with macroscopic, measurable quantities like the mass, charge, number and energy densities; bulk velocity; pressure and temperature. Particular attention is given to transformation of phase-space related coordinates and corresponding *transformations of the distribution function*. A whole section is dedicated to the *pressure tensors* and related definitions, and the physical meaning of their components, especially the perpendicular and parallel pressures in the case of a magnetically trapped ensemble of charged particles. In the definition of the perpendicular and parallel temperatures, we try to deactivate the usual tendency to associate these concepts with a Maxwellian distribution of particle velocities, showing that even a mono-energetic and mono-pitch angle ensemble of particles can have both. The example of a static distribution of trapped particles is examined in detail. Another section is dedicated to flux mapping in *invariant* Φ, J, M *space* and corresponding expressions, including their mutual transformations and relations to particle distribution functions in phase space and other invariant coordinate spaces. The chapter concludes with a brief discussion of trapped *particle diffusion*, with emphasis on the processes involved and their effects on the distribution function. The *Fokker-Planck equation* is derived for the radial diffusion case, and some general qualitative rules for the determination of *diffusion coefficients* are given.

The aim of Chap. 5 is to analyze and help understand the magnetospheric plasma as a self-organizing entity with self-generated electromagnetic fields (whereas in all previous chapters, the magnetic and electric fields were *given*, of sources external to the particle population). It starts with an introduction to *collisionless plasma physics* exclusively based on the understanding of adiabatic motion of individual particles. The concepts of *particle (kinetic) and guiding center fluids* are introduced as yet another example of "physics as the art of modeling". The corresponding distribution functions and their relations to macroscopic quantities are examined for the hypothetical case of identical particle ensembles, linking magnetization density with perpendicular pressure in a guiding center fluid. On the basis of very simple examples (throughout the text we call these "kindergarten examples"), the physical meanings of *equivalent and convective current densities* and their return circuits in a guiding center fluid are analyzed in detail, with emphasis on their origin in geometric aspects of cyclotron motion. Different classes of current densities are defined in general terms and their role in the generation of *magnetic stresses* in a particle ensemble is thoroughly examined. We finally turn to quasi-neutral mixtures of positive and negative particles, introducing the so-called *center of mass fluid*, discussing its properties and related equations. The concept of quasi-neutrality is examined and the plasma parameter known as *Debye length* is introduced; the reason why the magnetic field does not appear in the Debye length is discussed explicitly. All this leads to the plasma *momentum and magnetohydrodynamic equations*. The chapter concludes with the introduction of collisions and the formulation of the so-called *generalized Ohm equation*; the physical meaning of its terms is discussed as well its link to Maxwell's equations and the "chicken-and-egg" problem of whether currents drive fields or fields drive currents in a plasma. Several simplifications for some special situations are discussed, introducing concepts like *Hall conductivity*, *magnetic field diffusion*, *frozen-in magnetic field lines* and *Alfvén waves*.

To conclude this Preface and to facilitate the job of eventual book reviewers, we list some likely critical comments and give our pertinent replies/justifications:

1. *There are no problems, questions or exercises*. Correct. We believe that any instructor lecturing on this subject will be perfectly able to create meaningful problems which are not just mathematical "plug-in" exercises, and are tailored appropriately to the level of his/her class. Besides, throughout the text there are several standard statements such as "it is easy to show that ...". So just do that as an exercise!
2. *Where are the data?* Nowhere. For two reasons: the book is on theory and the really exciting new data are only now coming in.
3. *Why are magnetospheric models not discussed?* To show the fundamental physics of trapped particles, it is enough to deal with uncomplicated, unsophisticated models that still feature the most important characteristics of the real field, and which can even lead to analytical expressions.
4. *Where are the Euler coordinates?* In some footnotes. They are very useful tools for the math involved, but devoid of physical meaning (no direct, intuitive relation to the sources of the field).
5. *There are very few references, and my own work is not quoted.* True. But for instance in basic physics textbooks there usually are no references whenever Newton's or Maxwell's equations are used. Ours is a textbook, not a collection of review articles. Besides, there *are* some references—of papers which include abundant literature sources.
6. *The title of the book should be "Kinematics of Magnetically ... "*. Touché! But consider this: in any basic dynamics textbook, we find statements such as "let's apply a force *here*" or "consider a system of mass points *under the following*

constraints ...", with no words about the interaction processes responsible for those forces or constraints. And none are necessary if we just want to understand what happens and why.

Happy reading and, more importantly, happy understanding!

Fairbanks, Alaska, USA — Juan G. Roederer
August 2013 — Hui Zhang

References

1. J.A. Van Allen, G.H. Ludwig, E.C. Ray, C.E. McIlwain, Observations of high intensity radiation by satellites 1958 Alpha and Gamma. Jet Propuls. **28**, 588–592 (1958)
2. S.N. Vernov, A.E. Chudakov, E.V. Gorchakov, J.L. Logachev, P.V. Vakulov, Study of the cosmic-ray soft component by the 3rd Soviet Earth Satellite. Planet. Space Sci. **1**, 86 (1959)
3. D.N. Baker, S.G. Kanekal, V.C. Hoxie, M.G. Henderson, X. Li, H.E. Spence, S.R. Elkington, R.H.W. Friedel, J. Goldstein, M.K. Hudson, G.D. Reeves, R.M. Thorne, C.A. Kletzing, S.G. Claudepierre, A long-lived relativistic electron storage ring embedded in Earth's outer Van Allen belt. Science **340**(6129), 186–190 (2013)
4. G.D. Reeves, H.E. Spence, M.G. Henderson, S.K. Morley, H.W. Friedel, H.O. Funsten, D.N. Baker, S.G. Kanekal, J.B. Blake, J.F. Fennell, S.G. Claudepierre, R.M. Thorne, D.L. Turner, C.A. Kletzing, W.S. Kurth, B.A. Larsen, J.T. Niehof, Electron acceleration in the heart of the Van Allen radiation belts. Science **341**(6149), 991–994 (2013)
5. J.G. Roederer, *Dynamics of Geomagnetically Trapped Radiation* (Springer, New York, 1970)
6. T.G. Northrop, *The Adiabatic Motion of Charged Particles* (Interscience, New York, 1963)
7. B. Rossi, S. Olbert, *Introduction to the Physics of Space* (McGraw Hill, New York, 1970)
8. S. Lejosne, Modélisation du phénomène de diffusion radiale au sein des ceintures de radiation terrestres par technique de changement d'échelle. PhD thesis, Université de Toulouse, France, 2013

Contents

About the Authors

Juan G. Roederer Professor of Physics Emeritus, University of Alaska-Fairbanks. Born 1929 in Trieste, childhood in Vienna, higher education in Buenos Aires. Professor at the universities of Buenos Aires, Denver and Alaska. Former Director, Geophysical Institute, University of Alaska; former chairman, Advisory Committee on Earth and Space Sciences, Los Alamos National Laboratory, and the US Arctic Research Commission; former Senior Adviser, International Centre for Theoretical Physics, Trieste. Member of the Academies of Science of Austria and Argentina, and the World Academy of Sciences; Fellow of the American Association for the Advancement of Science and the American Geophysical Union. Recipient of the medal '100 Years of Geophysics' of the former Soviet Academy of Sciences, and of four NASA awards. His research areas range from cosmic rays and high energy physics to magnetospheric physics, psychophysics of music and information theory.

Hui Zhang Assistant Professor, Physics Department and Geophysical Institute, University of Alaska-Fairbanks. Born 1979 in Qingdao, China. She received her Bachelor of Science degree from Peking University, China, in 2002, and her Ph.D. in Astronomy from Boston University in 2008. She worked at NASA Goddard Space Flight Center for two years before moving to University of Alaska-Fairbanks in 2010, where at present she is in charge of the courses of elementary modern physics and fundamentals of plasma physics. Author or co-author of 39 peer-reviewed journal publications and presenter of 14 invited talks at national and international conferences and research institutes. Former guest editor of the *Journal of Atmospheric and Solar-Terrestrial Physics*, frequent reviewer of various journals and funding agencies. Co-chair of a Geospace Environment Modeling (GEM) focus group and convener/organizer of several international conferences. Her research area is space plasma physics with emphasis on solar wind-magnetosphere interactions.

Chapter 1
Particle Drifts and the First Adiabatic Invariant

1.1 Introduction: Adiabatic Theory and the Guiding Center Approximation

The equation which describes the motion of a particle of charge q and mass m in a magnetic field $\boldsymbol{B}$, under the action of an electric field $\boldsymbol{E}$ and an external non-electromagnetic force $\boldsymbol{F}$, is given by

$$\frac{d}{dt}\left(m\frac{d\boldsymbol{r}}{dt}\right) = q\left(\frac{d\boldsymbol{r}}{dt} \times \boldsymbol{B} + \boldsymbol{E}\right) + \boldsymbol{F} \text{ [1]} \tag{1.1}$$

The solution $\boldsymbol{r} = \boldsymbol{r}(t, \boldsymbol{r}_0, \boldsymbol{v}_0)$ represents the position of the particle as a function of time t, initial position $\boldsymbol{r}_0$ and initial velocity $\boldsymbol{v}_0$. Equation (1.1) is described from a given inertial frame of reference which we henceforth call the *original frame of reference* (OFR) (also called the laboratory system).

The solution of (1.1) may represent a very complicated trajectory. For instance, cosmic ray particle orbits in the geomagnetic field usually are of such nature. But under certain conditions of field geometry, external forces and particle energy like those prevalent in the radiation belts during geomagnetically quiet times, standard periodicities appear which then allow a simplified description if one is not interested in the actual instantaneous values of the phases in question, that is, if only *average* or *approximate* positions of the particles are wanted. There can be as many as three distinct types of periodicities: (i) the *cyclotron motion*, a periodicity in the particle's motion perpendicular to the magnetic field; (ii) the *bounce motion*, a periodic motion up and down a magnetic field line; and (iii) the *drift motion*, a periodic motion on a closed surface or drift shell, made up of field lines. Periodicity

[1] We shall use rationalized SI (Système International) units throughout this book. q is thus expressed in Coulombs (elementary charge $= 1.6021 \times 10^{-19}$ Coulombs), B in Tesla ($= 10^4$ Gauss)—see Appendix A.1.

J.G. Roederer and H. Zhang, *Dynamics of Magnetically Trapped Particles*, Astrophysics and Space Science Library 403, DOI 10.1007/978-3-642-41530-2_1,

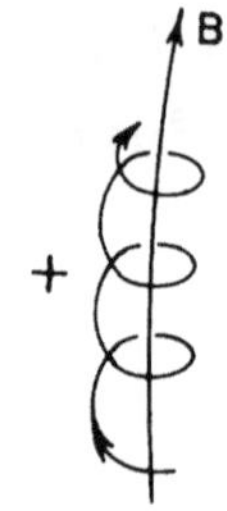

Fig. 1.1 Cyclotron motion along a field line

(i) is always the first to appear and has the highest frequency. It may exist even in absence of (ii) and (iii). For instance, when a not-too-high energy cosmic ray particle approaches the earth in the polar regions, it may have a clear-cut cyclotron motion but no other periodicity. Bounce frequencies are usually orders of magnitude lower than cyclotron frequencies; drift frequencies are orders of magnitude lower than bounce frequencies. Any magnetic field in which particles have the capability of bounce motion (ii) is called a *trapping* field. If, in addition, periodicity (iii) may occur, we say that it has a configuration of *stable trapping*.

Cyclotron motion qualitatively represents a helical motion of the particle around a field line (Fig. 1.1). In more precise terms, we say that cyclotron motion exists if at any instant of time we can find a (moving) frame of reference in which an observer sees the particle in a nearly circular periodic orbit, nearly perpendicular to the magnetic field (with single periodicity at least during a few cycles). If such a frame of reference can be found, we say that the *guiding center approximation* holds, and we call this particular frame of reference the *guiding center system* (GCS). The geometric center of the orbit is the *guiding center*; the (average) radius ρ_C is called *Larmor radius*, cyclotron radius or gyroradius. The period associated with the cyclotron motion (time to complete one turn) is called the *cyclotron period* τ_C. If no reference frame can be found in which the orbit is periodic and closed, there is no cyclotron motion. This happens with very high energy cosmic rays in the earth's field. Note carefully that these are rather qualitative descriptions, and that ρ_C and τ_C are quantities defined in the yet-to-be-determined GCS.

With the terms "nearly circular", "nearly periodic", "nearly perpendicular" we mean that any deviation from "exactly..." is very small in relation to the order of magnitude of the variable in question. Such approximations are typical for the adiabatic theory of particle motion, about which we may state in "kindergarten terms" that it provides correct answers only as long as "we don't look too close and are not expecting too detailed information". A more precise definition and description of the conditions under which adiabatic theory is useful will be given further below.

In the OFR we can picture the motion of a charged particle as the superposition of a displacement of its guiding center with a cyclotron rotation of the particle about

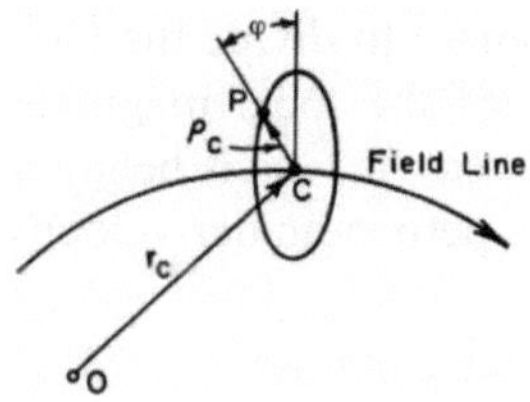

Fig. 1.2 Guiding center C, cyclotron radius ρ_C and phase angle φ

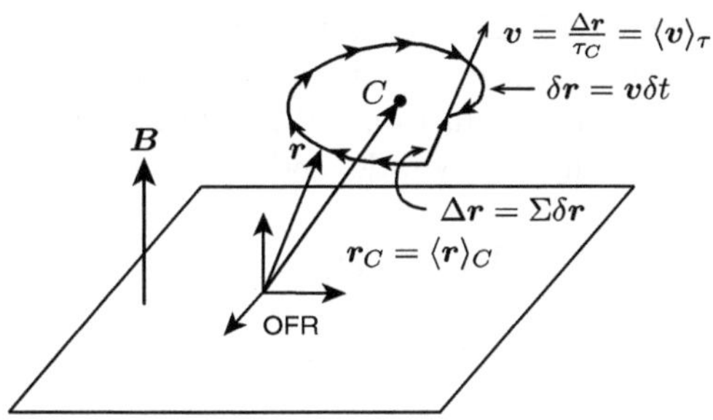

Fig. 1.3 Cyclotron orbit during one turn τ_C. The little orbital displacement vectors $\delta \boldsymbol{r} = \boldsymbol{v}\delta t$ add up to the vector $\Delta \boldsymbol{r}$ in one turn. $\Delta \boldsymbol{r}/\tau_C$ in the limit $\delta t \to 0$ defines the guiding center system velocity $\boldsymbol{V}$. The position $\boldsymbol{r}_C$ of the guiding center is defined as the cyclotron-average of the particle position vector $\boldsymbol{r}$ during one turn. In all this it is assumed that adiabatic conditions prevail, i.e., that $\Delta \boldsymbol{r} \ll \langle \boldsymbol{r} - \boldsymbol{r}_C \rangle$ and $\tau_C \ll \Delta t$, a characteristic time interval of "practical interest"

the guiding center. By definition, the instantaneous velocity of the guiding center is that of the GCS. The actual position $\boldsymbol{r}$ of the particle can thus be specified by the position $\boldsymbol{r}_C$ of the guiding center C, and a cyclotron *phase angle* φ in the GCS (provided the radius of gyration is known, Fig. 1.2). In many problems dealing with magnetically trapped particles, knowledge of the cyclotron phase is unimportant and the guiding center description is all that one needs. But forgetting φ completely would lead to nasty mistakes!

Considering Fig. 1.3, it is reasonable to define the instantaneous guiding center velocity vector $\boldsymbol{V}$ as the average of the particle's instantaneous velocity vector $\boldsymbol{v}$ in the OFR over a cyclotron period:

$$\boldsymbol{V} = \langle \boldsymbol{v} \rangle_\tau = 1/\tau_C \int_0^{\tau_C} \boldsymbol{v}\, dt \quad ^2 \qquad (1.2)$$

[2]For the time being we'll just consider τ_C as a priori known. As we shall see in the next section, in a non-relativistic situation τ_C is independent of the particle's dynamic state, depending only on mass, charge and the local magnetic field.

In adiabatic theory it is customary to divide the GC velocity into two components with respect to the direction of the local magnetic field (the natural coordinate system, Appendix A.1), each one of which behaves physically in a very distinct manner: $\boldsymbol{V} = \boldsymbol{V}_\perp + \boldsymbol{V}_\parallel$. The perpendicular velocity $\boldsymbol{V}_\perp$ of the GCS is called the particle's *drift velocity* $\boldsymbol{V}_D$:

$$\boldsymbol{V}_D = \boldsymbol{V}_\perp = \langle \boldsymbol{v}_\perp \rangle_\tau \tag{1.3}$$

because it represents the velocity with which we see the particle "drift away" from the initial field line in the OFR. Likewise, the parallel velocity $\boldsymbol{V}_\parallel$ is the average of the particle's instantaneous parallel velocity $\boldsymbol{v}_\parallel$. However, contrary to what happens with vector $\boldsymbol{v}_\perp$, whose direction turns a full 2π during one cyclotron period, the parallel velocity vector changes very little, and we can write

$$\boldsymbol{V}_\parallel = \langle \boldsymbol{v}_\parallel \rangle_\tau \cong \boldsymbol{v}_\parallel \tag{1.4}$$

Quite generally, the cyclotron-averaging symbol $\langle \ldots \rangle_\tau$ represents the integral operator $1/\tau_C \int_0^{\tau_C} \ldots dt$.[3]

Finally, the position vector of the guiding center is the cyclotron average of the particle's position vector (Figs. 1.2 and 1.3):

$$\boldsymbol{r}_{GC} = \langle \boldsymbol{r} \rangle_\tau \tag{1.5}$$

In the GCS (starred quantities), the particle's instantaneous velocities parallel and perpendicular to the magnetic field are, respectively,

$$\begin{aligned} \boldsymbol{v}^*{}_\parallel &= 0 \\ \boldsymbol{v}^*{}_\perp &= \boldsymbol{v}_\perp - \boldsymbol{V}_D \end{aligned} \tag{1.6}$$

By definition, for the particle velocity vector in the GCS:

$$\langle \boldsymbol{v}^*{}_\perp \rangle_\tau = 0 \tag{1.7}$$

For the moduli:

$$\begin{aligned} v^*{}_\perp{}^2 &= v_\perp{}^2 - 2\boldsymbol{v}_\perp \cdot \boldsymbol{V}_D + V_D{}^2 \\ \langle v^*{}_\perp{}^2 \rangle_\tau &= \langle v_\perp{}^2 \rangle_\tau - V_D{}^2 \end{aligned} \tag{1.8}$$

[3] Later we will run into the *cyclotron phase* average operator, $\langle \ldots \rangle_\varphi = 1/2\pi \int_0^{2\pi} \ldots d\varphi = 1/\tau_C \int_0^{\tau_C} \ldots (d\varphi/dt) dt$. When the cyclotron motion in the GCS is uniform, both operators are identical and will be designated as $\langle\ \rangle_C = \langle\ \rangle_\tau = \langle\ \rangle_\varphi$. A compilation of all phase averages used in this book is given in a footnote on page 186 of Appendix A.3.

The latter equation tells us that $0 \leq \langle {v^*_\perp}^2 \rangle_\tau \leq \langle {v_\perp}^2 \rangle_\tau$. In particular, when the magnitudes $\langle {v_\perp}^2 \rangle_\tau$ and $\langle {v^*_\perp}^2 \rangle_\tau$ coincide, $V_D = 0$, i.e., there is no drift. Less trivially, when $\langle {v_\perp}^2 \rangle_\tau = {V_D}^2$ we have $v^*_\perp \equiv 0$ and there is no cyclotron motion: the particle follows, at least locally, an "uncurled" trajectory in the OFR (an example will be discussed in the next section).

For the magnetic and electric fields in a moving GCS the following transformations apply:

$$\begin{aligned} \boldsymbol{B}^* &= \boldsymbol{B} \\ \boldsymbol{E}^* &= \boldsymbol{E} + \boldsymbol{V} \times \boldsymbol{B} \end{aligned} \tag{1.9}$$

Notice that only the perpendicular part $\boldsymbol{V}_D$ of the GCS velocity $\boldsymbol{V}$ contributes to the induced electric field term in (1.9). We have used non-relativistic transformations in anticipation of the fact that, for all practical radiation belt and plasma configurations $V_D \ll c$ (velocity of light)—even if the particles themselves may be relativistic.

Expression (1.9) is particularly important. It indicates that for any given point in a $\boldsymbol{B}, \boldsymbol{E}$ field one can always find a moving frame of reference for which the component of $\boldsymbol{E}$ perpendicular to the local $\boldsymbol{B}$ has been "transformed away", i.e., in which $\boldsymbol{E}^*_\perp = 0$ at that point (if both $\boldsymbol{B}$ and $\boldsymbol{E}$ are uniform over a finite domain, one can find one common frame for all points therein). This fact plays a fundamental role both in adiabatic theory and plasma physics.

The velocity $\boldsymbol{U}$ of this special frame at point $\boldsymbol{r}$ (in the OFR) and time t can be obtained by multiplying vectorially the second equation of (1.9) by $\boldsymbol{B}$:

$$\boldsymbol{U}(\boldsymbol{r}, t) = \frac{\boldsymbol{E} \times \boldsymbol{B}}{B^2} \tag{1.10}$$

$$U(\boldsymbol{r}, t) = \frac{E_\perp}{B} \tag{1.11}$$

Note that, again, only the perpendicular component $E_\perp$ intervenes in these relations. $E_\parallel$ survives transformation (1.9) intact.[4]

There are different types of drifts (1.3), which appear under different well-defined circumstances. They can be classified into various groups according to whether they depend on the dynamic variables of the particle or according to the restrictive conditions that have to be imposed to guarantee their validity. We shall discuss each group independently on the basis of particular examples.

[4] In regions where $B \to 0$ (e.g., near a neutral line) the concept of "transforming away $E_\perp$" breaks down. See page 18.

1.2 Uniform Magnetic Field; Basic Definitions; Magnetic Moment

As the most basic example we consider a charged particle in a *uniform*, static magnetic field, in absence of any external forces. We rewrite (1.1) in the form:

$$\frac{d\boldsymbol{p}}{dt} = q\boldsymbol{v} \times \boldsymbol{B} \tag{1.12}$$

where $\boldsymbol{p}$ is the particle's momentum, and $\boldsymbol{v}$ its velocity in the OFR. The right hand side is called the *Lorentz force*; it is always perpendicular to the particle's velocity. Therefore, in absence of non-magnetic forces, the speed v and the kinetic energy T remain constant. We can write (1.12) in the form

$$m\boldsymbol{a} = q\boldsymbol{v} \times \boldsymbol{B} \tag{1.13}$$

which also holds relativistically because $m = \text{const}$ in this particular case. Also $\boldsymbol{B}$ is constant in space and time. The angle between $\boldsymbol{v}$ and $\boldsymbol{B}$

$$\alpha = \arccos\frac{v_{\parallel}}{v} = \arcsin\frac{v_{\perp}}{v} \tag{1.14}$$

is called the particle's *pitch angle*. For the parallel component of (1.13) we have:

$$m\boldsymbol{a}_{\parallel} = q|\boldsymbol{v} \times \boldsymbol{B}|_{\parallel} \equiv 0$$

This means the motion of the particle projected along a uniform magnetic field is rectilinear uniform. Since $|\boldsymbol{v}| = \text{const}$, we also conclude that

$$v_{\perp} = \text{const.} \tag{1.15}$$

$$\alpha = \text{const.} \tag{1.16}$$

For the perpendicular component of (1.12) we can write $m\boldsymbol{a}_{\perp} = q|\boldsymbol{v} \times \boldsymbol{B}|_{\perp}$. The acceleration $\boldsymbol{a}_{\perp}$ is therefore always perpendicular to $\boldsymbol{v}_{\perp}$ and its magnitude

$$a_{\perp} = \frac{q}{m}v_{\perp}B = \text{const.} \tag{1.17}$$

in view of (1.15).[5]

[5]Henceforth, whenever the charge q appears in the expression of a *scalar* quantity as $a_{\perp}$, it will be meant to represent the *absolute value* of q (unless explicitly stated to the contrary). If on the other hand q appears in the expression of a *vector* quantity, it is assumed *to carry its actual sign*; otherwise it will be explicitly written as $|q|$.

This means that the particle's motion projected on a plane perpendicular to the magnetic field is *circular uniform* and $a_\perp$ is the centripetal acceleration.

In the case under discussion, the GCS moves along the magnetic field with velocity

$$\boldsymbol{V}_\parallel = \boldsymbol{v}_\parallel = \text{const. and } \boldsymbol{V}_\perp = 0 \tag{1.18}$$

In the GCS, $v_\perp = $ const and the particle trajectory is a circle with a gyroradius that can be obtained from (1.17):

$$\rho_C = \frac{v_\perp^{*\,2}}{a_\perp} = \frac{m v_\perp^*}{qB} = \frac{p_\perp^*}{qB} \tag{1.19}$$

We have used the starred quantity $v_\perp^*$ to re-emphasize the fact that the gyroradius ρ_C is defined in the GCS (in the particular case under discussion, though, $v^* = v$ because $V_D \equiv 0$.) Notice that, in view of the factor q in (1.13), positive and negative particles have mutually opposite senses in their cyclotron rotations.

Associated with the particle's cyclotron motion, we have the cyclotron period (defined in the GCS like ρ_C):

$$\tau_C = \frac{2\pi\rho_C}{v^*{}_\perp} = \frac{2\pi m}{qB} \tag{1.20}$$

and the cyclotron angular frequency

$$\omega_C = \frac{2\pi}{\tau_C} = \frac{qB}{m} \tag{1.21}$$

Expressions (1.19)–(1.21) are valid relativistically, provided one considers m as the relativistic mass $m = m_0\gamma$ (m_0 rest mass; $\gamma = (1-\beta^2)^{-1/2}$; $\beta = v/c$). In the non-relativistic case, ω_C and τ_C are *independent of the particle's velocity*, depending only on the field intensity and the class of particles (q/m); in other words they are a function of space (a scalar field). This fully justifies the definition of the drift velocity (1.2) as an average over cycle time. Using (1.14), it is useful to express (1.19) and (1.20) in the form:

$$\rho_C = \frac{m_0 c}{q}\frac{1}{B}\beta\gamma\sin\alpha \tag{1.22}$$

$$\tau_C = \frac{2\pi m_0}{q}\frac{1}{B}\gamma \tag{1.23}$$

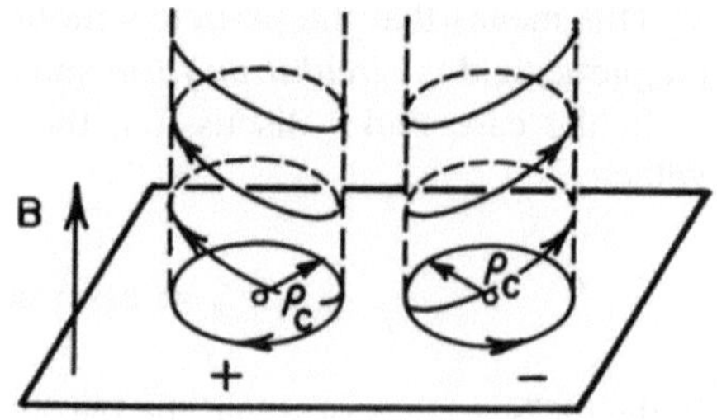

Fig. 1.4 Positive and negative particles in a uniform field

For electrons and protons, respectively, the constant factors have values:

$$\frac{m_0 c}{q} = 1.705 \times 10^{-3} \text{ and } 3.13 \text{ (Tesla·m)}$$

$$\frac{2\pi m_0}{q} = 3.57 \times 10^{-11} \text{and } 6.56 \times 10^{-8} \text{ (Tesla·s)}$$

From now on we shall deal mainly with non-relativistic ($\gamma = 1$) cases. It is possible to express the gyroradius in vector form taking into account (1.19) and (1.21):

$$\boldsymbol{\rho}_C = -\frac{m}{qB^2}\boldsymbol{v}^* \times \boldsymbol{B} = -\frac{1}{\omega_C}\boldsymbol{v}^* \times \boldsymbol{e} \tag{1.24}$$

where $\boldsymbol{e} = \boldsymbol{B}/B$ is a unit vector in the direction of $\boldsymbol{B}$ (see Appendix A.1). The vector $\boldsymbol{\rho}_C$ points from the guiding center to the particle . The instantaneous position of the guiding center of a particle of velocity $\boldsymbol{v}$ at point $\boldsymbol{r}$ thus becomes

$$\boldsymbol{r}_{GC} = \boldsymbol{r} + \frac{m}{qB^2}\boldsymbol{v}^* \times \boldsymbol{B} = \boldsymbol{r} + \frac{1}{\omega_C}\boldsymbol{v}^* \times \boldsymbol{e} \tag{1.25}$$

Since during one cyclotron turn the vector $\langle \boldsymbol{v}^* \rangle_\tau = 0$ (1.6), we have $\boldsymbol{r}_{GC} = \langle \boldsymbol{r} \rangle_\tau$, as defined earlier.

The motion of a charged particle in a uniform magnetic field is circular helicoidal. Positive particles spiral clockwise around the field lines if we look at them in a direction opposite to $\boldsymbol{B}$; negative particles spiral counter-clockwise (Fig. 1.4). When the pitch angle is 90°, $\boldsymbol{v}_\parallel = 0$ and the motion is perpendicular to $\boldsymbol{B}$; the particle has cyclotron motion only and stays on a circle forever. In that case, the GCS coincides with the OFR. If $\alpha = 0$, there is no cyclotron motion at all; the particle moves along a straight field line. Notice that in general ρ_C is *not* the radius of curvature of the particle's trajectory in the OFR (this is true only if $\alpha = \pi/2$). For a helicoid, the radius of curvature R_C is always greater than the radius of the cylinder around which the helicoid is wound ($R_C = \rho_C / \sin^2 \alpha = \rho_C (v/v_\perp)^2$).

As viewed from the GCS, the cyclotron motion of a charged particle about the guiding center is equivalent to a circular electric current loop with radius ρ_C and intensity $I = |q|/\tau_C = q^2 B/(2\pi m)$ of the same direction for both positive and

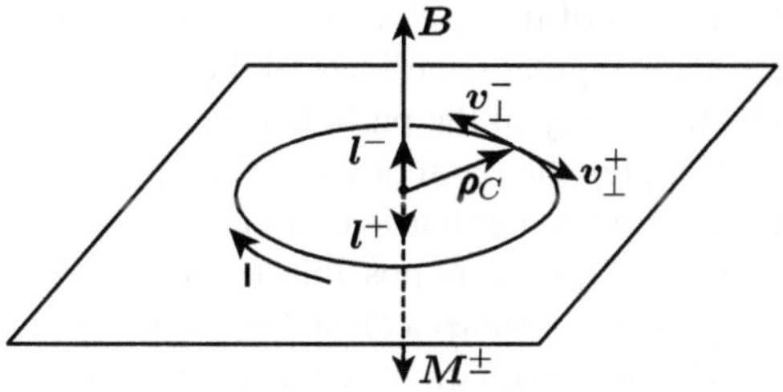

Fig. 1.5 Magnetic moment $\boldsymbol{M}$ and angular momentum $\boldsymbol{l}$ of positive and negative particles

negative particles, Fig. 1.5.[6] The associated magnetic moment $\boldsymbol{M} = I\delta\boldsymbol{S}$ is directed opposite to $\boldsymbol{B}$, of magnitude $M = I\pi\rho_C{}^2 = mv_\perp^{*\,2}/2B = p_\perp^{*\,2}/(2mB)$ (1.19). Thus we can write:

$$\boldsymbol{M} = -\frac{mv_\perp^{*\,2}}{2B}\frac{\boldsymbol{B}}{B} = -\frac{p_\perp^{*\,2}}{2mB}\boldsymbol{e} = -\frac{T^*{}_\perp}{B}\boldsymbol{e}$$

$$M = \frac{mv_\perp^{*\,2}}{2B} = \frac{p_\perp^{*\,2}}{2mB} = \frac{T^*{}_\perp}{B} \tag{1.26}$$

The expression of the transverse kinetic energy $T^*{}_\perp = 1/2mv_\perp^{*\,2}$ is valid only for the non-relativistic case. It is important to emphasize that $v_\perp^*$ is the modulus of the transverse velocity in the GCS.

In its cyclotron motion, the particle also has an angular momentum or spin about the guiding center $\boldsymbol{l} = m\boldsymbol{\rho}_C \times \boldsymbol{v}^*$ directed opposite to $\boldsymbol{B}$ for a positive charge and in the same direction for negative particles (Fig. 1.5). Taking into account (1.19) and (1.26) we can write:

$$\boldsymbol{l} = -\frac{m^2 v_\perp^{*\,2}}{qB}\boldsymbol{e} = 2\frac{m}{q}\boldsymbol{M} \tag{1.27}$$

Although we have defined (1.26) and (1.27) for the case of a uniform $\boldsymbol{B}$ field in absence of other forces, they are of general validity and, indeed, of crucial importance: M and l are constants of motion within the guiding center approximation, i.e., *adiabatic invariants.* In other words, the magnetic moment M and the spin l are intrinsic parameters associated with a particle in cyclotron motion. This is the reason why in the Introduction we talked about replacing the original particle with a virtual particle, normally called “guiding center particle” or “magnetized charged particle” of the same mass m and charge q, but in which the “averaged-out” cyclotron motion is represented by just one vector: the magnetic moment $\boldsymbol{M}$ (or the intrinsic angular momentum $\boldsymbol{l}$) which we mentally picture as being attached to the GC particle. What we lose in this “remodeling” process are

[6] Yet another model is to imagine the charge q smeared evenly over the cyclotron circle in the GCS, rotating uniformly with period τ_C. See page 23.

the details of the cyclotron motion, specifically, the cyclotron phase of the original particle (angular coordinate of the particle in its cyclotron motion, Fig. 1.2). The full kinetic energy of the particle can be retrieved from the value of M and the local B (1.26), plus the parallel GCS velocity $V_{\parallel}$. The overall condition for the validity of the guiding center particle model is that the requirements for the GC approximation apply, i.e., that it is possible to identify a frame of reference (the GCS) in which the particle executes a closed-orbit periodic motion (for a more precise definition of the GC approximation, see next section). In the case of Fig. 1.4, a GC particle just moves straight up or down along the magnetic field line with a velocity $\boldsymbol{V}_{\parallel}$ equal to the parallel velocity of the parent particle.

It is important to note that for the external field's magnetic flux Φ through the cyclotron orbit in the GCS (see (1.26) and (1.27)),

$$|\Phi| = \pi {\rho_C}^2 B = \pi (m^2 {v_\perp^*}^2 / q^2 B) = (2\pi m/q^2) M = (\pi/q) l \qquad (1.28)$$

Consequently, Φ is also an adiabatic constant of motion.

The expression (1.28) can be used for a "kindergarten"-level demonstration of the *conservation of the magnetic moment M* (non-relativistic case). Suppose that in the field configuration under discussion (90° particles in a uniform B-field), the magnetic field intensity changes in time very slowly, so that $\tau_C \ll B/(dB/dt)$. According to Faraday's law, the induced electric field $\boldsymbol{E}_I$ will change the transverse kinetic energy of the particle during one cyclotron turn by $\delta T_\perp^* = q \oint \boldsymbol{E}_I \boldsymbol{ds} = q\, d\Phi/dt$ (absolute values only). To first order (in which we consider ρ_C constant), the change per unit time will be (first equality in (1.28)): $dT_\perp/dt \simeq \delta T_\perp^*/\tau_C = q/\tau_C\,(d\Phi/dt) = T_\perp^*/B\,(dB/dt)$—therefore, $T_\perp^*/B = M = \text{const}$. In essence, this means that in the above cyclotron loop model, the current I is a closed current with the well-known property of a superconducting system to react (driven by the induced electric field) to any change of the magnetic field in such a way so as to maintain constant the flux Φ through its own loop. Thus a cycling particle will adjust its cyclotron orbit so as to preserve Φ, hence its magnetic moment M (as long as the variation of the magnetic field satisfies the adiabatic condition). Although we have shown this for the simplest magnetic field configuration possible, it has general validity: the particle doesn't care *why* the flux through its cyclotron orbit changes—only that it *does*![7]

When the particle velocity is relativistic, it can be demonstrated that the quantity that is conserved is the *relativistic magnetic moment*

[7]In Hamiltonian mechanics (e.g., [1]) of point charges in a magnetic field, it is demonstrated that for cyclic variables like the arc $\boldsymbol{l}$ (see Fig. 1.5) the so-called canonical path or action integral $J = \oint (\boldsymbol{p} + q\boldsymbol{A}) \cdot \boldsymbol{dl}$ is a constant of motion for a single particle (provided that the fields and the forces change very little during one cycle). Taking $\boldsymbol{dl}$ in the direction of a positive particle (Fig. 1.5) and carefully considering that, therefore, the magnetic flux through the cyclotron loop $\oint \boldsymbol{A} \cdot \boldsymbol{dl}$ is negative, we have $J_c = \oint (\boldsymbol{p} + q\boldsymbol{A}) \cdot \boldsymbol{dl} = 2\pi \rho_C m v_\perp^* - \pi {\rho_C}^2 B = (2\pi m/q) M$, therefore $M = \text{const}$.

$$M_r = \frac{p_\perp^{*\,2}}{2m_0 B} = \frac{1+\gamma}{2}\frac{T_{r\perp}^*}{B} \tag{1.29}$$

where m_0 is the rest mass and $T_{r\perp}^*$ the perpendicular relativistic kinetic energy ($T_r = m_0c^2(\gamma - 1) = m_0v^2\gamma^2/(1+\gamma)$). Since the rest of the book deals mainly with non-relativistic particles, some of the principal equations can be easily converted to relativistic ones by replacing the magnetic moment, wherever it appears, with its relativistic expression (1.29) and the kinetic energy with the relativistic kinetic energy T_r (for a full relativistic treatment of adiabatic theory, see [2, 3] which also shows the full relativistic expressions of the most important relationships).

Speaking of relativity, it is interesting to briefly examine the cyclotron motion of a charged particle from the quantum physics point of view. The stronger the magnetic field, for a given transverse velocity of the particle the higher will be its cyclotron frequency (1.21) and the smaller its gyroradius (1.19). For ultra-high intensity fields, quantum mechanics must be applied in the description of the cyclotron motion; this situation is of importance in the theoretical study of particles trapped in the magnetic field of a neutron star or black hole. The magnetic field intensities in the environment of such an object may be as high as $10^6 - 10^{11}$ Tesla. Consider Heisenberg's uncertainty relation $\Delta x \Delta p_x \geq \hbar/2$ ($\hbar$ = Planck's constant$/2\pi = 1.05 \times 10^{-26}$ J). If as a meaningful maximum order of magnitude we insert for Δx the gyroradius (1.17) and for Δp_x the product $mv_\perp$, taking into account (1.27) the uncertainty relation gives a lower limit for the GC particle's angular momentum: $l \geq \hbar/2$. In other words, zero is not an option and the intrinsic angular momentum of a GC particle in a magnetic field is quantized. As a matter of fact, the quantum energy levels (called Landau levels, [4]) of an elementary charged particle gyrating in an intense magnetic field B turn out to be

$$E_n = \frac{q\hbar}{m_0}(n + \frac{1}{2})B \qquad n = 0, 1, 2, \ldots$$

This leads to energy levels of the order of hundreds of keV to thousands of MeV, for electrons trapped in extreme magnetic environments. It also means that the cyclotron frequency is quantized:

$$\omega_n = \frac{qB}{m_0}(n + \frac{1}{2}) \qquad n = 0, 1, 2, \ldots$$

Considerations of quantum spin of the original particle and relativistic effects complicate somewhat the picture, but are outside the scope of this book.

1.3 Zero-Order Drifts

We now consider the case of a charged particle in a uniform static magnetic field under the action of an external non-magnetic, non-inertial, interaction force $\boldsymbol{F}$ which is constant in time and space. We divide the force into two components $\boldsymbol{F}_{\parallel}$ and $\boldsymbol{F}_{\perp}$, parallel and perpendicular to $\boldsymbol{B}$, respectively. The equation of motion (1.1) can be split into the following pair:

$$\frac{d\boldsymbol{p}_{\parallel}}{dt} = \boldsymbol{F}_{\parallel}$$

$$\frac{d\boldsymbol{p}_{\perp}}{dt} = \boldsymbol{F}_{\perp} + q\boldsymbol{v}_{\perp} \times \boldsymbol{B}$$

The first equation tells us that the particle is accelerated along the field line in a "conventional" way by $\boldsymbol{F}_{\parallel}$. Let us assume for the time being that $\boldsymbol{F}_{\parallel} = 0$ (i.e. $\boldsymbol{F} = \boldsymbol{F}_{\perp}$). This means that the particle will have a constant velocity $v_{\parallel}$ along the field line and so will the GCS. Thus we only need to determine the GCS's perpendicular velocity or drift velocity, which we call $\boldsymbol{V}_F$. This is the instantaneous velocity of a frame of reference in which the particle executes a circular motion. To find $\boldsymbol{V}_F$, we will use the "trick" of creating a motion-induced electric field $\boldsymbol{E}^*$ (1.9) in a moving frame of reference such that the external force is balanced out by a force $q\boldsymbol{E}^* = q\boldsymbol{V}_F \times \boldsymbol{B}$:

$$q\boldsymbol{E}^* + \boldsymbol{F}^*{}_{\perp} = q\boldsymbol{V}_F \times \boldsymbol{B} + \boldsymbol{F}_{\perp} = 0 \tag{1.30}$$

Multiplying vectorially by $\boldsymbol{e}/qB$, where $\boldsymbol{e}$ is again the unit vector in the direction of $\boldsymbol{B}$, we obtain

$$\boldsymbol{V}_F = \frac{\boldsymbol{F}_{\perp} \times \boldsymbol{e}}{qB} = \frac{\boldsymbol{F} \times \boldsymbol{e}}{qB} \tag{1.31}$$

$\boldsymbol{V}_F$ is called the *force drift*. Since $\boldsymbol{F}_{\parallel}$ does not affect (1.31), this expression is also valid in the more general case when $\boldsymbol{F}_{\parallel} \neq 0$.

As viewed from the OFR, the particle has a cyclotron motion (that in the GCS) plus a translation with constant velocity $\boldsymbol{V}_F$ given by (1.31), plus a translation parallel to the field line. If both $\boldsymbol{F}_{\parallel}$ and $\boldsymbol{v}_{\parallel}$ are zero, the resulting motion is a cycloid in a plane perpendicular to B (Fig. 1.6). Notice that $\boldsymbol{V}_F$ is always perpendicular to both $\boldsymbol{B}$ and $\boldsymbol{F}$. The particle thus "reacts" perpendicularly to the external force and *no average work is done* on the particle during its drift motion under the present condition of a uniform field, although in the OFR, the kinetic energy of the particle changes periodically in its cyclotron turns; in the GCS, however, the perpendicular velocity is $v_{\perp}^* = \text{const}$. Positive and negative particles drift in mutually opposite directions. Most importantly, $\boldsymbol{V}_F$ is *independent of the particles' mass and energy*. It is called a *zero order drift* because the only condition for its validity is that the

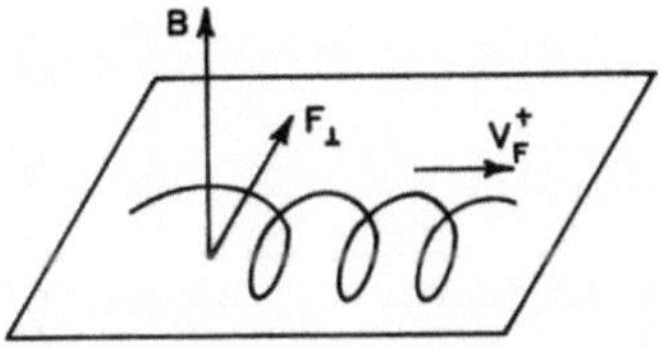

Fig. 1.6 Drift $\boldsymbol{V}_F^+$ in a homogenous magnetic field $\boldsymbol{B}$ under a force $\boldsymbol{F}_\perp$

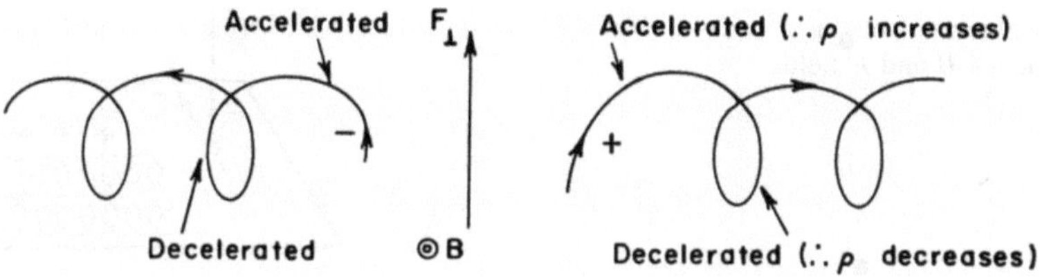

Fig. 1.7 Physical cause for the existence of a force drift

extension of the uniform field domain be large enough to allow the particle to execute its cyclotron or spiral turns undisturbed (domain $\gg \rho_C$). Note that for a force field, zero order drifts are a function of position $\boldsymbol{r}$ only—a vector field. We cannot independently impart some arbitrary zero order drift to a particle as an initial condition.

One can easily understand the physical reasons for the drift of a charged particle under the action of a constant external force, perpendicular to $\boldsymbol{B}$. As sketched in Fig. 1.7, the kinetic energy of a particle in the OFR is *not* constant; during one cycloid turn, the particle is alternatively being accelerated (larger radius of curvature ρ—do *not* confuse with Larmor radius!) and decelerated by the external force $\boldsymbol{F}$—$v_\perp$ is not constant during one turn (see below). It should be clear that in this case the magnetic moment cannot be defined in the OFR.

If the external force is not constant but derives from a general force field $\boldsymbol{F} = -\nabla W$ (W: force field potential), (1.31) still is valid. Let us assume that $F_\parallel = 0$ everywhere, which means that the magnetic field lines lie in electrostatic equipotential surfaces. The force-drift velocity (1.31) is tangent to such an equipotential surface and the guiding center of a 90° pitch angle particle will thus *follow an equipotential line* in a plane perpendicular to the uniform field $\boldsymbol{B}$ (Fig. 1.8). This, however is true only under the assumption that the variation of W over a gyroradius is very small with respect to the particle's kinetic energy:

$$\Delta W_{\rho C} \ll T \tag{1.32}$$

This in turn implies that

$$V_F = \frac{F}{qB} \ll v_\perp \tag{1.33}$$

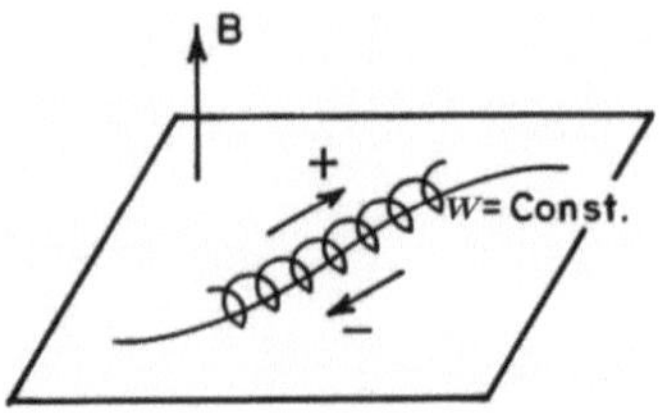

Fig. 1.8 Force drifts along a force-equipotential line

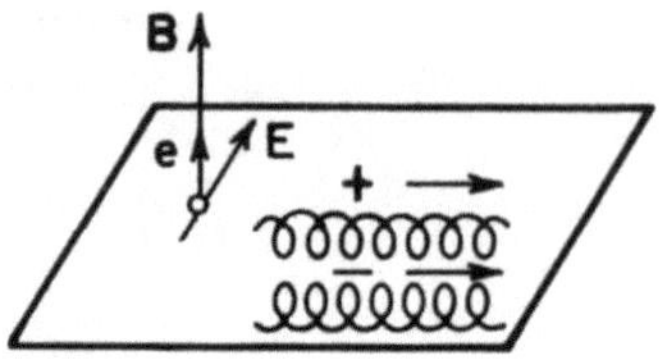

Fig. 1.9 Electric drifts in homogeneous B and E fields

The particle will execute many overlapping cyclotron turns as it drifts along a $W = \text{const}$ curve.

The most important kind of external force is the electric field force $q\boldsymbol{E}$. The only non-electric external force of interest would be gravitation; it plays a role for plasmas in stellar environments but none in planetary magnetospheres. However, as we shall see later, there are inertial forces of importance (virtual non-interaction-based forces arising in an accelerated frame of reference). With an electric field, the drift velocity (1.31) becomes

$$\boldsymbol{V}_E = \frac{\boldsymbol{E} \times \boldsymbol{B}}{B^2} = \boldsymbol{U} \tag{1.34}$$

This is called the "*E-cross-B drift*", for which we shall always use the letter U. Note the crucial fact that the electric charge has canceled out; both positive and negative particles drift in the same direction with the same speed, regardless of their mass and energy (Fig. 1.9). This common electric drift $\boldsymbol{U}$ is identical to the velocity (1.10) of a frame of reference in which the perpendicular component $\boldsymbol{E}_\perp$ has been transformed away. It should come as no surprise, for in our case this is, indeed, the frame of reference in which there are no forces other than the Lorentz force acting on the particle—the very definition of the GCS! The fact that $\boldsymbol{U}$ is a drift velocity common to all particles confirms its fundamental role as the bulk velocity of the ensemble of particles that constitute a plasma. In contrast, the charge-dependent force drift (1.31), and other drifts which we shall introduce later, give rise to electric currents in a plasma (Chap. 5).

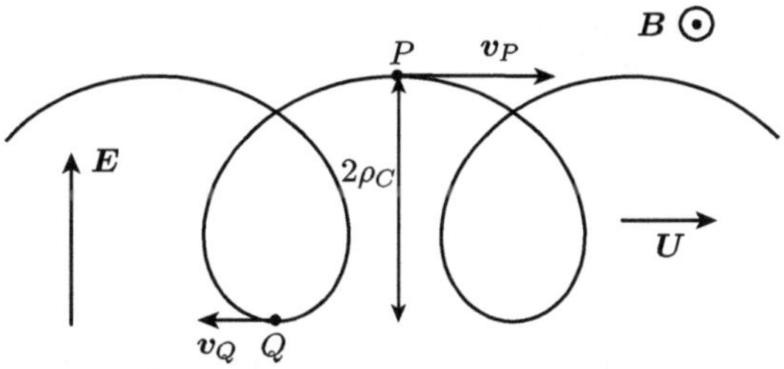

Fig. 1.10 OFR parameters for an "E cross B" drift

1.4 Examples of "*E* Cross *B*" Drifts; Uniform Magnetic Field of Time-Dependent Intensity

It is instructive to examine the motion of a charged particle in uniform $\boldsymbol{B}$ & $\boldsymbol{E}$ fields as viewed "under a magnifying glass" in the OFR. Consider Fig. 1.10. The uniform magnetic field is directed out of the paper, and $v_{\parallel} = 0$. We assume only the direction of $\boldsymbol{U}$ as known, but not its modulus. Call v_P and v_Q the velocities along the x axis of the particle at points P and Q, respectively (consider one of these velocities as given). We will have, from (1.6) and with $v^* = v^*{}_{\perp} =$ const.:

$$v_P = v^* + U$$
$$v_Q = v^* - U$$

On the other hand, we have the following energy relation:

$$\frac{1}{2}m(v_P{}^2 - v_Q{}^2) = qE\,2\rho_C$$

Therefore, with (1.19)

$$U = \frac{E}{B}$$

which is identical to (1.11). But in this case we have not used in any way the transformation (1.9).

The gyroradius ρ_C (1.19) can be expressed as a function of v_P or v_Q:

$$\rho_C = \frac{m}{qB}\left|v_P - \frac{E}{B}\right| = \frac{m}{qB}\left|v_Q + \frac{E}{B}\right|$$

Note that this is *not* the radius of curvature of the trajectory in the OFR!

It is interesting to compare the total forces acting in the OFR on the particle at P and Q, respectively:

$$|f_{P,Q}| = qv_{P,Q}B \mp qE = qv^*B$$

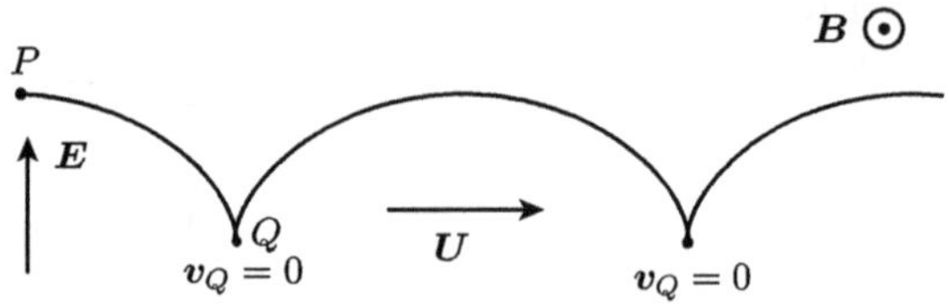

Fig. 1.11 Special case of an "*E* cross *B*" drift

Both relations represent the constant magnitude of the Lorentz force in the GCS. The different radii of curvature of the orbit at P and Q in the OFR arise from the difference in particle velocities at these points (1.35):

$$R_{P,Q} = \frac{mv_{P,Q}^2}{f_{P,Q}} = \frac{m(v^* \pm U)^2}{qv^*B} = \rho_C\left(1 \pm \frac{U}{v^*}\right)^2$$

Notice that if $U = E/B = 1/2\,v_P$, we have $v^* = U$ and $R_Q = 0$; the trajectory is the limit of an open cycloid (Fig. 1.11). This shows that a particle can be instantaneously *at rest* in the OFR and yet possess cyclotron and drift motion![8] On the other hand, if $U = E/B = v_P$, we have $v^* = 0$ and $R_Q = \infty$; the trajectory in the OFR is a straight line (in the GCS the particle is at rest!). In this case, in the OFR, the Lorentz force is balanced out by the electric field force at all times (the principle of a velocity spectrometer!).

Returning to the kinetic energy, in the GCS the particle's T^* is constant, as is its speed v^*. The cyclotron-average kinetic energy transverse to $\boldsymbol{B}$ in the OFR will be $\langle T_\perp \rangle = 1/2m\langle v^2{}_\perp \rangle$. Taking into account that $\boldsymbol{v}_\perp = \boldsymbol{v}_\perp^* + \boldsymbol{U}$ (we reintroduce the subindex $\perp$ because what follows is valid in general, also for $v_\parallel \neq 0$), and that by definition (see (1.3) and (1.6)) $\langle \boldsymbol{v}_\perp \rangle = \boldsymbol{U}$ and $\langle \boldsymbol{v}_\perp^* \rangle = 0$, we have

$$\langle T_\perp \rangle = T_\perp^* + \frac{1}{2}mU^2 \tag{1.35}$$

To consolidate the understanding of the electric drift velocity (1.34) (answering for instance the question: When $B \to 0$, what does it really mean that the guiding center races away with $V_E \to \infty$?) we discuss another simple but illustrative example of possible detailed configurations of charged particle trajectories in a uniform $\boldsymbol{B}\&\boldsymbol{E}$ field. Consider the case of positive particles injected one after another with the same initial velocity $\boldsymbol{v}_i$, as shown in Fig. 1.12 ($\boldsymbol{B} \perp \boldsymbol{E} \perp \boldsymbol{v}$). While each particle is traveling, B (and E) are held constant; initially, the magnetic field is large. Then B is decreased before the next injection. This process is repeated until B reaches zero.

[8] If we place a charged particle in a $B\&E$ field with *zero* initial velocity (for instance, by ionizing a neutral atom at rest), it will start moving in an open cycloid (Fig. 1.11), with a drift velocity U given by (1.10) and a Larmor radius $\rho_C = mE/qB^2$. The maximum kinetic energy of the particle (at point P) will be $T = 2m(E/B)^2 = 2mU^2$, and the average energy, according to (1.35), will be mU^2. This so-called "ion pick-up" process plays an important role in space physics.

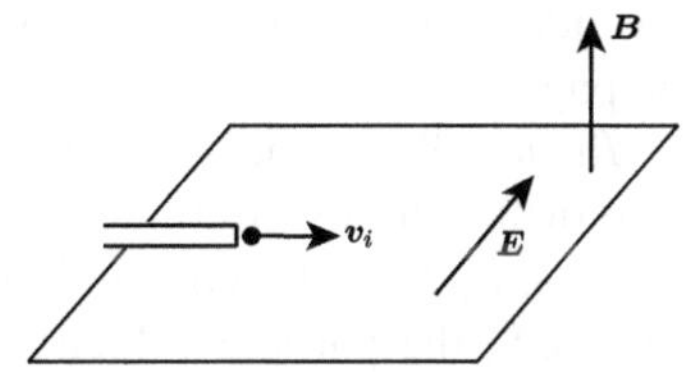

Fig. 1.12 Set-up for particles injected with constant velocity v_i into a $\boldsymbol{E} \perp \boldsymbol{B}$ field, in which $E =$ const. and B is gradually being decreased to zero

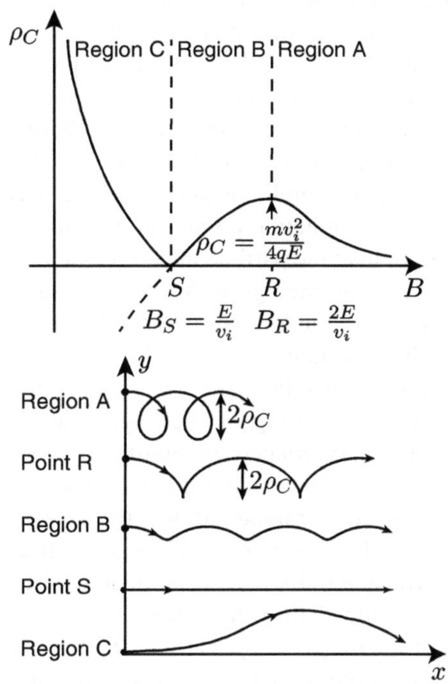

Fig. 1.13 *Top*: Larmor radius ρ_C as a function of $|B|$ for the example of a charged particle of given initial velocity $\boldsymbol{v}_i$ injected perpendicularly into uniform $\boldsymbol{B}$ and $\boldsymbol{E}$ fields. Definition of characteristic regions and points discussed in the text. *Bottom*: Sketch of typical orbits in the OFR for characteristic regions and points. Not in scale!

Let us analyze in detail the cyclotron and drift motions of the particles, as they are injected into smaller and smaller magnetic fields. From (1.35) we have

$$\rho_C = \frac{m}{qB}\left|v_i - \frac{E}{B}\right| \tag{1.36}$$

(bars because $\rho_C > 0$ always!). Figure 1.13 (top) shows the graph of the function $\rho_C = \rho_C(B)$.

Consider the distinct regions A, B and C, and the "notable points" R and S, where $B_R = 2E/v_i$ and $B_S = E/v_i$, respectively.

Region A: For large B's the drift velocity (1.10) and the Larmor radius (1.19) are very small and the particle behaves adiabatically, turning many times before drifting away (see Fig. 1.13, lower graph). As the B decreases, the gyroradius and the drift velocity U increase. In the GCS, the particle velocity v^* decreases.

Point R: The Larmor radius reaches a maximum; in the GCS the velocity $v^* = U$ and in the OFR the particle trajectory becomes the limit of an open cycloid, with cuspidal points at which the particle comes momentaneously to rest (see footnote on page 16).

Region B: As the B field decreases further, the gyroradius begins to decrease. The drift velocity continues to increase and the trajectory is an open cycloid with increasing "wavelength" and amplitude—the drift U wins over the velocity v_i. In the GCS, the particle velocity v^* continues to decrease, and the cyclotron circles become smaller and smaller.

Point S: In the GCS the cyclotron circle has been reduced to a point: the particle is at rest, $v^* = 0$! In the OFR the particle moves in uniform rectilinear motion with speed U (the velocity spectrometer effect mentioned above).

Region C: As the magnetic field intensity continues to decrease, the drift speed increases further, and in the OFR the particle follows an open cycloid with decreasing amplitude and wavelength. However now it is a cycloid turning *upwards* from the injection point, into the direction of $\boldsymbol{E}$. In this regime the GC velocity is larger than the particle injection velocity; the GCS "races away" from the particle to the right in the figure. An observer in the GCS will in turn see the particle initially moving to the left, being at the lowest point of a circular motion.

The domain $c \lesssim E/B \lesssim \infty$ deserves special attention. Would this really represent a relativistic or "transrelativistic" situation for the GCS? What is the physical meaning of such apparent nonsense? Obviously we have a breakdown of the very concept of guiding center. Let us not forget that the GC is a purely *geometric* feature and that the "GC particle" is a *virtual* artifact, a useful model, product of our imagination—the only physical reality is the original cycling particle! The drift velocity $\boldsymbol{U}$ is also a geometric concept, mathematically tied to the original particle. All this can be summarized by stating that the restriction for the validity of zero-order drifts is that the spatial domain of the $\boldsymbol{B}$ & $\boldsymbol{E}$ field be much larger than the gyroradius ρ_C of the particle. More specifically, if L_{trans} and L_p are the extensions of the field domain transverse and parallel to the particle drift $\boldsymbol{U}$, respectively, the conditions for adiabatic behavior in a uniform E-cross-B field are

$$\rho_C \ll L \quad \text{or} \quad v \ll \frac{qB}{m} L_{trans}$$

$$N_{turns} \gg 1 \quad \text{or} \quad U = \frac{E}{B} \ll \frac{qB}{2\pi m} L_p \tag{1.37}$$

In the second equation N_{turns} is the number of cyclotron turns when the particle drifts the length L_p. When according to (1.36) ρ_C becomes large enough but the conditions of validity still hold, the particle could, indeed, be accelerated by the E field to relativistic velocities before the Lorentz force wins and the particle turns over and enters a deceleration phase. For a B-value strictly zero, we simply have the case of a particle under the action of just one force ($q\boldsymbol{E}$), and the trajectory becomes a *parabola* in the non-relativistic domain.

The example discussed above shows that what must be used in the expression of M (1.26) and l (1.27) is the perpendicular velocity v^* of the particle in the GCS, not in the OFR where the transverse velocity of the particle $v_\perp$ is variable. For instance, in case S above ($B = E/v_i$), $\rho_C = 0$, and M and l are both zero! On the other hand, a particle injected into a $B\&E$ field with zero initial velocity *does* have a non-zero magnetic moment $M = mE^2/2B^3$ (the "ion pick-up mechanism" mentioned in the footnote on page 16—best argument yet that the transverse velocity in the definition (1.26) of M is *not* the velocity in the OFR!). It is important to emphasize again that zero order drifts are independent of the particle's energy as long as the physical domain of uniform fields is large enough (1.37).

Our next example of $\boldsymbol{E}$ cross $\boldsymbol{B}$ drift, of particular conceptual importance in magnetospheric plasma physics, is that of the drift of a guiding center particle in a purely *induced* electric field (time-dependent magnetic field, no charges present anywhere)

$$\boldsymbol{V}_{ind} = \boldsymbol{E}_{ind} \times \frac{\boldsymbol{B}}{B^2} = -\frac{\partial \boldsymbol{A}}{\partial t} \times \frac{\nabla \times \boldsymbol{A}}{|\nabla \times \boldsymbol{A}|^2} \tag{1.38}$$

(see Appendix A.1). The first obvious case should be that of a stationary uniform magnetic field, whose source current system moves rigidly with constant velocity $\boldsymbol{V}_0$ with respect to a frame of reference at rest (the OFR). No electric charges are present. A 90° pitch angle charged particle would be subjected to an induced electric field drift velocity given by (A.57) of Appendix A.1; in other words, it would drift with the moving frame velocity and thus obviously remain part of the magnetic field system. Notice that this time we have appealed to the vector potential and Maxwell's equations rather than simply using the postulated rule of field transformation (1.9). Although we only considered a uniform magnetic field in pure translation here, the same procedure applies to rigidly *rotating* fields like the internal geomagnetic and planetary fields (neglecting external magnetospheric currents): at each point there will be an induced electric field and associated zero-order drift velocity (1.38), which in this case is called *corotational drift*, causing all trapped particles to corotate with the planet. Of course, additional drifts (see next sections) will complicate the picture. In the case of Earth, the corotational electric field will have a significant effect only on very low energy particles (e.g., the constituents of the plasmasphere), but in the magnetospheres of Giant Planets it plays a very important role even for relativistic electrons of their radiation belts.

As another example of induced electric field drift, we consider a long circular solenoid with a time-dependent current (Fig. 1.14). We again shall assume the scalar potential to be zero (no free electric charges present); the vector potential will be that of relation (A.56) of Appendix A.1, directed as shown in the figure and of magnitude $A = B\,r$. We now inject a 90° pitch angle particle of energy or magnetic moment such that $\rho_C \ll R$, radius of the solenoid. If nothing else happens, its guiding center will remain at rest. Now we adiabatically increase the current in the solenoid, such that $\dot{B} > 0$. An induced electric field will appear as shown in Fig. 1.14, given by (A.58). Independently of its mass, charge and energy, the particle will drift toward the center of the solenoid with an induced drift speed of $V_{ind} = \dot{B}\,r$. Due to the conservation of magnetic moment (1.26), its energy will change at a rate $dT/dt = M\dot{B}$—i.e., in this example, gradually increase. A whole ensemble

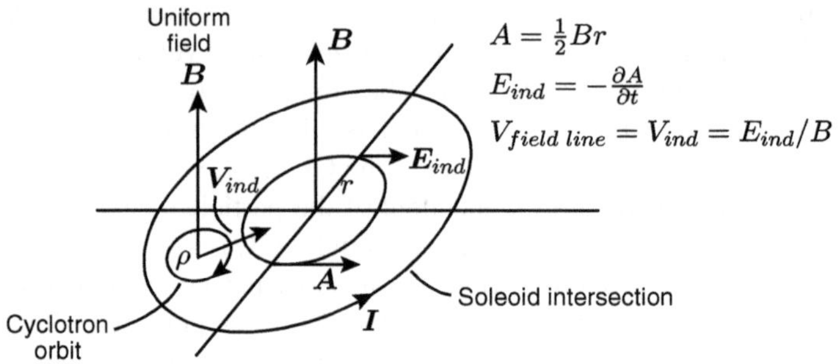

Fig. 1.14 Configuration of the vector potential $\boldsymbol{A}$ and the induced electric field $\boldsymbol{E}_{ind}$ in a solenoid field which decreases in time

of particles trapped in the increasing uniform field would thus be energized and compressed toward the axis.

This example allows us to introduce the concept of *field line motion* (e.g., [5]) in a phenomenological way. Because the preceding process affects all $\alpha = 90°$ charged particles equally regardless of their nature, if we "paint" a given initial field line by placing such particles all along it, they will all drift together and remain on a common field line (see first corollary at the end of Appendix A.2) in a way governed *exclusively* by what happens to the magnetic field (its sources) time-wise. It is thus possible to *declare* the "painted" field line to be the same as the initial one and adopt (1.38) as the ad hoc operational definition of *velocity of the point of a field line*. Although we only have considered the oversimplified case of a uniform magnetic field, it can be shown that the property of remaining on a common field line applies to any changing magnetic field configuration—as long as there are *no potential* electric fields ($\nabla \cdot \boldsymbol{E} \equiv 0$ everywhere) and *no parallel* electric fields are present ($\partial \boldsymbol{A}/\partial t \cdot \boldsymbol{B} \equiv 0$).[9] For the general case, we also must add the condition of near-zero energy ($T \to 0$) for the probe particle, to avoid the action of other drifts (see following sections), which are all energy-dependent. Note that the above ad hoc

[9]Without the first condition we would run into the undesirable situation of having "moving field lines" in a static magnetic field crossed by a static electric field, as for instance in the case of two oppositely charged plates placed parallel to $\boldsymbol{B}$ in the gap of a magnet. It can be shown that the above definition is not only independent of the particular position of the guiding center along the field line but that it is magnetic flux-preserving (i.e., flux tubes preserve their identity). There is nothing artificial with a definition using probe particles placed along a field line: after all, as we know from elementary textbooks, the electric and magnetic field vectors themselves are formally defined by forces on probe particles! Neither is the condition that all ρ be turned off artificial: in elementary electromagnetism books, the self-field of probe charges are also "turned off" (ignored). While it is tempting to exaggerate the physical significance of a purely mathematical-geometric concept such as a field line, we must never lose sight of the fact that the only "physical reality" in electromagnetism is mutually interacting electrically *charged matter*. Here are some perhaps too strong words from Richard Feynman: "…not only is it not possible to say whether field lines move or do not move with charges—they may disappear completely in certain coordinate frames …" (see [6]).

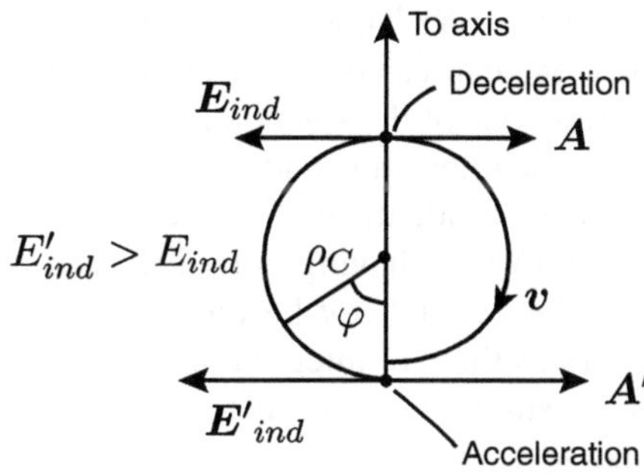

Fig. 1.15 Single cyclotron orbit dynamics in a gradually increasing B-field

definition of moving field line is valid even in a vacuum situation; note also that thus defined, it is always perpendicular to the field line in question—a parallel field line velocity cannot be defined in a physically meaningful way (although we may well imagine it, for instance, for the field of a long solenoid being transported parallel to its axis). The reason for using terms like "to declare" and "ad hoc" is related to the fact that, as mentioned in Appendix A.1, a *physical* distinction between the potential and inductive contributions to an electric field as required prior to use of (1.38) is not possible through a single measurement operation (we would "have to know" the system a priori, or probe it through an experimental protocol by turning off all electrostatic sources).

Finally, we take advantage of the above very simple example to present a better "microscopic" view of the cyclotron acceleration (or deceleration) process ultimately responsible for the conservation of a particle's magnetic moment under adiabatic conditions. Looking with a magnifying glass at the particle in Fig. 1.14 in the OFR, we see it gyrating in an induced electric field more explicitly shown in Fig. 1.15. Notice that since the $\boldsymbol{A}$-vector increases linearly with r, the particle will be alternatively accelerated and decelerated, with the acceleration phase always winning a bit in our example. This causes *both* the drift toward the axis and the gradual increase in kinetic energy, and is why, in reality, the orbit shown in Fig. 1.15 is *not* closed in the OFR. Now we can come up with a more convincing analytical, less kindergarten-like, proof of the conservation of magnetic moment—albeit only for a very simple field configuration. In the figure and according to the discussion on page 176 of Appendix A.1, the magnitude of the induced electric field will be $E_{ind} = \dot{A} = \dot{B}(r + \rho_C \cos\varphi)$, where φ is the cyclotron phase as measured from the lower point of the orbit. In addition to the near-constant transverse Lorentz force $q\boldsymbol{v}\times\boldsymbol{B}$, in the case of a varying magnetic field there will be a *tangential* electric force $f_t = q\,\dot{B}(r + \rho_C \cos\varphi)\cos\varphi$, varying from positive to negative during one half-turn. Although under adiabatic condition this tangential force will be much smaller in magnitude than the Lorentz force, the near-circular motion of the gyrating particle is not uniform. The work of this tangential force during a half-turn, i.e., the kinetic energy gain of the particle, turns out $\Delta T = \int_0^{\pi} f_t \rho_C \, d\varphi = q\dot{B}\rho_C{}^2\,\pi/2$. Dividing by $\tau_C/2$ we obtain the change of kinetic energy per unit time of the gyrating particle in a time dependent (increasing) uniform field: $\dot{T} = 2\Delta T/\tau_C = 1/2mv^2\,(\dot{B}/B)$ or

$(\dot{T}/T) = (\dot{B}/B)$, which if integrated gives $T/B = M = \text{const.}$ (1.26). In a similar way one can easily figure out the induced drift of the gyrating particle toward the central axis of the solenoid.

As an interesting addendum to this proof we note that if at time t there are several particles of identical velocities on a *common* cyclotron orbit like the one shown in Fig. 1.15 (i.e., having only different phases), a simple dynamic calculation shows that they would *not* be on common orbits at later times *except* at times t_n after intervals that are integer multiples of the cyclotron period: $t_n = t + n\,\tau_C$. We will find a similar behavior in the case of other periodicities of adiabatic motion (bounce and drift).

1.5 First Order Drifts

We now drop the assumption of a uniform magnetic field. Gradients in the magnetic field (assumed constant in time) cause first order drifts which are energy-dependent. A magnetic field is inhomogeneous when any of the components of the gradient tensor $\partial B_i/\partial x_k$ (Appendix A.1) are non-zero in some region of space. There are different kinds of field gradients from the geometrical point of view which play quite distinct roles in adiabatic theory. For the time being, we work only with the *vector* gradient of the *modulus* B, which we denote by ∇B. Its components are of course related to those of the tensor gradient (see relation (A.35) of Appendix A.1). We divide this vector into two components, parallel and perpendicular to the magnetic field, respectively (see Appendix A.1, (A.20) and (A.19)):

$$\nabla B = \nabla_{\perp} B + \nabla_{\parallel} B$$

Each component has a distinct effect on the guiding center motion.

In this section we examine the effect of the perpendicular gradient $\nabla_{\perp} B$. In Appendix A.1 (A.21) and (A.15) it is shown that in a current-free region in which $\nabla_{\perp} B \neq 0$, all field lines are curved; in spite of this, we shall ignore field line curvature for the time being (as we shall see in Sect. 1.6 this is fully justified for 90° pitch angle particles in the equatorial (minimum-B) surface of a trapping magnetic field). We shall also assume condition (1.39) to apply: the magnetic field varies very little over the Larmor radius of the particle. If ∇B is the modulus of the gradient of B and $B/\nabla B$ is a characteristic length for the change of B, the *adiabatic condition* can be written in the following equivalent ways:

$$\rho_C \ll \frac{B}{\nabla_{\perp} B} \quad ; \quad \nabla_{\perp} B \ll \frac{|q| B^2}{m v_{\perp}} \tag{1.39}$$

Let us inject a 90° pitch angle particle with velocity $\boldsymbol{v}_{\perp}$ into a field with (almost) straight field lines and a perpendicular gradient (Fig. 1.16). We realize

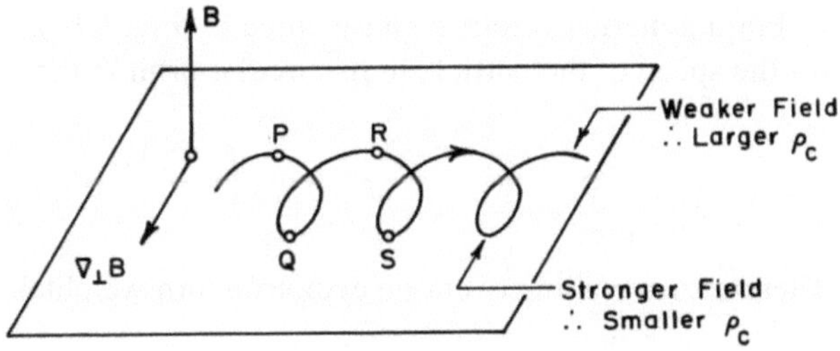

Fig. 1.16 Showing the physical reasons for the gradient-B drift of a positive particle

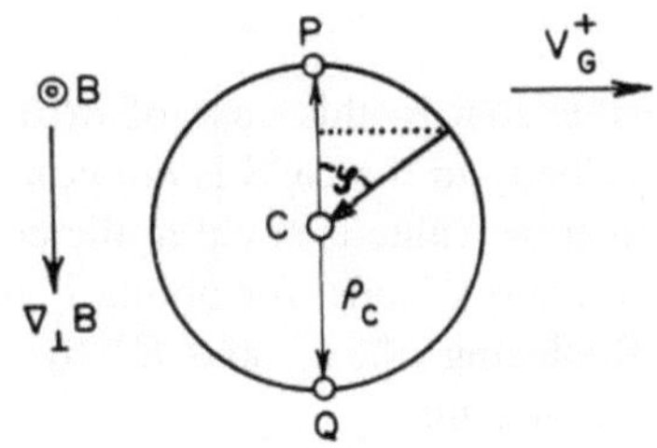

Fig. 1.17 Cyclotron orbit in the GCS in a non-uniform field

that there must be a drift to the right: for a positive particle the field at points P, R, …is weaker than that at points Q, S, …, causing an alternating change of the radius of curvature of the particle orbit (note carefully that the orbit in the figure is drawn "stretched-out"; because of condition (1.39), the particle really turns many times before appreciably drifting away from the initial position). We now want to find the general expression for the corresponding *gradient-B drift velocity* $\boldsymbol{V}_G$, i.e., the perpendicular velocity of the GCS. First, we note that the particle must move with constant speed $v_\perp$ *in the OFR* (rather than in the GCS as happens with force drifts), since there are no external non-magnetic forces acting. As we did in the case of a zero order drift in a uniform $B\&E$ field, we shall use the motion-induced electric field trick. Let us assume that the GCS indeed moves to the right with velocity $\boldsymbol{V}_G$ as intuitively shown in Fig. 1.16. The particle velocity in the GCS is $\boldsymbol{v}^*$ (we drop the obvious subindex $\perp$); but now the motion-induced electric field $\boldsymbol{E}^* = -\boldsymbol{V}_G \times \boldsymbol{B}$ (1.9) will alternatively accelerate and decelerate the particle. Therefore, its speed in the GCS will not be constant: the motion in the GCS is circular but non-uniform (reverse situation from the force drift case in the precious section!) In particular, for points P and Q in Fig. 1.17:

$$\begin{aligned} v^*_{P,Q} &= v_\perp \mp V_G \\ B_{P,Q} &= B \mp \rho_C \nabla_\perp B \qquad (1.40) \\ E^* &= V_G B \end{aligned}$$

B is the magnetic field at the guiding center C.

For a generic cyclotron phase angle φ (Fig. 1.17), and taking into account (1.6) for the speed of the particle in its circular orbit in the GCS,

$$v^*(\varphi) = v_\perp \left(1 - \frac{V_G}{v_\perp} \sin\varphi\right) \tag{1.41}$$

Therefore, averaging over one cyclotron turn we obtain

$$\langle v^* \rangle = v_\perp \tag{1.42}$$

Note that in this case of drift in an inhomogeneous magnetic field, the particle velocity in the GCS is *not* constant (despite the orbit being a closed circle), but its average value is equal to the constant speed in the OFR. In the GCS, conservation of energy leads, for points P and Q in Fig. 1.16, to $\frac{1}{2}m({v_P^*}^2 - {v_Q^*}^2) = 2\rho_C q E^*$. Replacing v_P^*, v_Q^* and E^* by their expressions in (1.40), we find that V_G cancels out, leaving

$$\rho_C = \frac{m v_\perp}{qB} \tag{1.43}$$

Notice that in this case, what enters in the expression of the Larmor radius is *not* the varying velocity in the GCS, but its *average* value, which in this case happens to be equal to the constant speed in the OFR. The same happens with the magnetic moment. Defined in the GCS, we really must use in its fundamental expression (1.26) the *average* $\langle v_\perp^* \rangle$. Does this invalidate the derivation of this expression based on a simple circuit model? No, we just have to improve a bit that model: instead of one particle of charge q circling with a velocity $v_\perp^*$ we must "smear" the charge along the circle with a linear charge density $\lambda = q/(2\pi\rho_C)$. The current I at any given point will be $I = \lambda v_\perp^*$. If the velocity varies along the circuit, as happens in the case under discussion, conservation of charge requires that $I = \lambda v_\perp^* =$ const. along the circle. Since the resulting magnetic moment only cares about I and not how the latter is made up as a convection current, the model circuit remains intact! Confused? Such nitpicking details do happen in adiabatic theory! It also shows how every model in physics must be constantly and carefully examined in detail regarding its validity and the degree of its approximation to the "reality out there".

We continue our detailed examination considering Newton's equation at points P and Q, involving centripetal accelerations:

$$m a_{P,Q} = \frac{m(v_\perp \mp V_G)^2}{\rho_C} = q v_\perp (B \mp \rho_C \nabla_\perp B) \tag{1.44}$$

The last terms are the Lorentz forces in the OFR at points P and Q, respectively. From either of the second equalities we obtain, inserting the above expression for ρ_C:

$$\left(1 \mp \frac{V_G}{v_\perp}\right)^2 = 1 \mp \frac{mv_\perp}{qB^2}\nabla_\perp B$$

or, finally, for the moduli:

$$V_G = \frac{mv_\perp{}^2}{2qB^2}\nabla_\perp B = \frac{T_\perp}{qB^2}\nabla_\perp B = \frac{M}{q}\frac{\nabla_\perp B}{B} \tag{1.45}$$

With (1.26) and (1.27), and observing Fig. 1.16, we obtain several versions for the *vector* expression of the gradient-B drift velocity:

$$\begin{aligned} \boldsymbol{V}_G &= \frac{1}{2}\frac{mv^2{}_\perp}{qB^3}\boldsymbol{B} \times \boldsymbol{\nabla}_\perp B \\ &= \frac{T_\perp}{qB^2}\boldsymbol{e} \times \boldsymbol{\nabla}_\perp B = -\frac{1}{q}\boldsymbol{M} \times \frac{\boldsymbol{\nabla}_\perp B}{B} = \frac{\boldsymbol{l}}{2m} \times \frac{\boldsymbol{\nabla}_\perp B}{B} \end{aligned} \tag{1.46}$$

These expressions are non-relativistic; for relativistic particles, we have for the magnitude of V_G:

$$V_G = \frac{m_0c^2}{2q}\beta_\perp{}^2\gamma\frac{\nabla_\perp B}{B^2} \tag{1.47}$$

in which

$$\frac{m_0c^2}{2q} = 2.56 \times 10^5 \text{ Tesla } m^2/\text{s for electrons}$$

$$\frac{m_0c^2}{2q} = 4.70 \times 10^8 \text{ Tesla } m^2/\text{s for protons}$$

Although the above expression is valid for relativistic particles, the actual values of the drift velocity in typical magnetospheric fields are non-relativistic; as we shall see, this is also the case for other drifts. The first equalities in (1.46) tell us that given a surface perpendicular to the magnetic field, a 90° pitch angle particle on it will drift along a $B =$ const. contour. As we shall see in the next section, this is an important result for magnetospheric physics.

Relations (1.46) were proven here for the very restricted case of a 90° pitch angle particle in a non-uniform magnetic field with nearly parallel field lines. They have general validity, provided condition (1.39) holds for the magnetic field gradient. Note that in contrast to the zero order drift (1.34), the gradient-B drift velocity depends on the energy, mass and charge of the particle and that, according to (1.45) and (1.43), the adiabatic condition (1.39) leads to

$$V_G \ll v_\perp \tag{1.48}$$

We now complete the "microscopic" examination of this case by returning to the OFR to calculate the different radii of curvature R_P and R_Q at points P and Q in

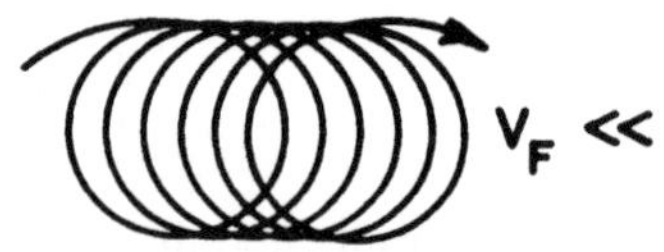

Fig. 1.18 Gradient-B drift obeying the adiabatic condition

Fig. 1.16—a difference that is responsible for the shape of the cycloidal orbit in the OFR. Taking the expressions for the Lorentz force at these points (last equalities in (1.44)), and equating them to the centripetal accelerations in the OFR (times mass), we obtain

$$R_{P,Q} = \rho_C \left(1 \pm \rho_C \frac{\nabla_\perp B}{B} \right)$$

If condition (1.39) holds, the difference between R_P and R_Q is, indeed, extremely small, and the cycloid will be tightly closed (Fig. 1.18). As an important remark, we note the following: if we determine the vector expression of the centripetal Lorentz force $\boldsymbol{f}(\varphi)_L$ in the GCS as a function of the phase angle (as we did with the velocity $\boldsymbol{v}^*$ in (1.41)), we would discover that the gradient-B drift (1.46) is nothing but a *force drift* caused by the phase-average force $\langle \boldsymbol{f}(\boldsymbol{\varphi})_L \rangle_\varphi$. It does not count as a zero-order drift, however, because the phase-average force is a first-order quantity, it is not energy independent and $V_G \ll v_\perp$ always. We will return to this in detail in the next chapter.

Let us summarize some of the principal properties of zero-order and first-order drifts, for non-relativistic, 90° pitch angle particles. In the *zero-order drift* due to the action of an external force, the drift of the particle is independent of its energy; the orbit in the original frame of reference is in general a cycloid and the particle velocity is variable (periodic). In the guiding center frame of reference, the orbit is circular uniform. The constant particle speed in the GCS is equal to the average of the particle speed in the OFR. There are no restrictions on particle energy, as long as there is enough space for the particle to execute its periodic motion. In a *first-order drift* due to a transverse magnetic field gradient, the drift motion depends on the particle energy; the orbit in the OFR is also a cycloid, but the speed is constant. In the GCS, the orbit is circular, but the speed is variable (periodic). The average speed in the GCS is equal to the constant speed in the OFR. There are restrictions on energy and field gradient.

1.6 Example: Drift of 90° Pitch-Angle Particles in the Magnetospheric Equator; Effects of an Electric Field

We shall now analyze the motion of 90° pitch angle particles on the equatorial (minimum-B) surface of the earth's magnetosphere making some simplifying assumptions about magnetic and electrostatic field models. The study of equatorial

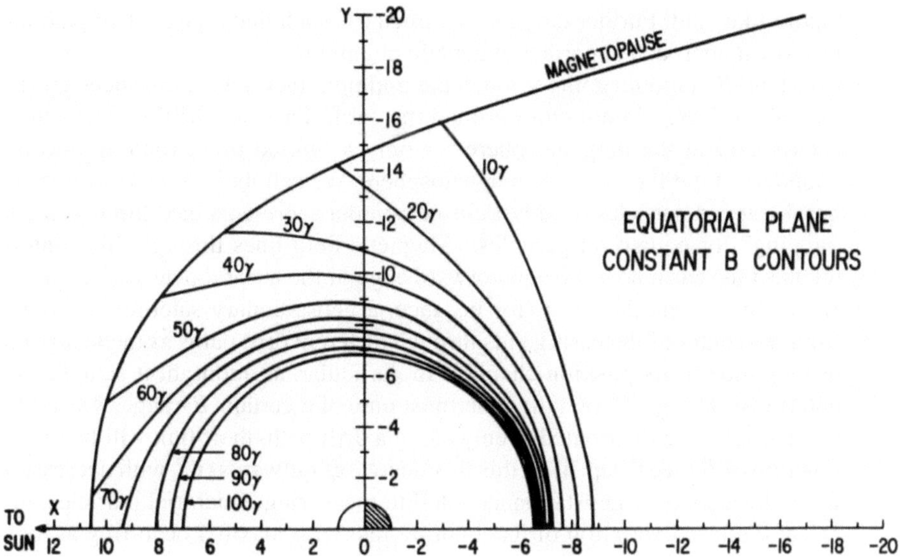

Fig. 1.19 Experimentally determined average contours of constant magnetic field intensity in the equatorial surface [9]. These contours represent drift paths for energetic, 90° pitch angle particles. Radial distances are in earth radii (1 $R_E = 6{,}371$ km)

particles is useful for several reasons: (i) as we shall see later, the equatorial point of a field line represents an equilibrium position for mirroring particles—hence the study of equatorial particles provides "first order" information on the behavior of the off-equatorial population; (ii) the theoretical treatment of equatorial and near-equatorial particles can be done analytically by using simple field models, providing physical insight (though not quantitative accuracy) into fundamental aspects of trapped particle dynamics [7, 8]; (iii) many characteristic effects of spatial field asymmetries are most pronounced for equatorial particles; (iv) more experimental information is available on trapped particles at low geomagnetic latitudes, especially in the outer magnetosphere.

Let us consider the drift motion of charged particles and assume that no external forces are acting (this is nearly the case for radiation belt electrons and protons of energies greater than about 100 keV). The guiding centers of these particles will experience a pure gradient drift (1.46) following constant-B curves; electrons will drift eastwards, protons westwards. Figure 1.19 sketches the principal features of constant magnetic field intensity contours as observed in systematic measurements [9]. A qualitative examination of the figure leads to the following conclusions:

(i) Within about 7∼8 earth radii (1 $R_E = 6{,}371$ km) all drift paths are closed. In that region an equatorial particle thus remains *stably trapped* (assuming that there are no external perturbations.)
(ii) The trajectories' day-night asymmetry increases as one moves away from the earth. Within about 4 R_E, they are approximately circles, as prescribed by a

dipole-like field. Further out, the asymmetry is such that a given drift path has its closest approach to earth at magnetic midnight.

(iii) Constant-B contours going through the midnight meridian at distances greater than about 7 R_E do not close around the earth. Particles drifting along them are trapped in the magnetosphere for only a limited time, running into the boundary at the flanks of the magnetosphere. We call them *pseudo-trapped* or quasi-trapped particles. The last closed contour represents the "limit of stable trapping" for equatorial particles. (Magnetic field lines through this contour intersect the earth near the equatorward edge of the auroral oval.)

(iv) A satellite in circular orbit (for instance, a geostationary satellite at 6.6 R_E from the center of the earth), cuts through different drift paths as the local time or longitude of its position changes. In particular, at midnight it samples the outermost B-ring, at noon the innermost one, of a certain B-range. Assuming that particles are distributed evenly along a drift path, their flux will be only a function of B (see Chap. 4). If this flux decreases outwards (i.e. with decreasing B), a detector on a geostationary satellite measuring equatorial particles will reveal a diurnal variation of its counting rate with maxima occurring at local noon.

(v) Drift paths are closer to each other at midnight. This means that the transverse gradient of the field is larger there. Thus, according to (1.46), a particle's drift velocity will be larger at night than at noon. As a consequence, stably trapped equatorial particles spend more time on the day side than on the night side during their drift. This difference becomes more pronounced as we approach the limit of stable trapping.

An analytical expression of the equatorial magnetospheric field intensity B_0 at point r_0, ϕ_0 (longitude east of midnight), reasonably good during quiet and moderately disturbed times in the region 1.5–7 R_E, is given:

$$B_0 = B_E\left(\frac{R_E}{r_0}\right)^3\left[1 + \frac{b_1}{B_E}\left(\frac{r_0}{R_E}\right)^3 - \frac{b_2}{B_E}\left(\frac{r_0}{R_E}\right)^4 \cos\phi_0\right] \tag{1.49}$$

The three terms represent, respectively, (i) the main dipole field (with $B_E \sim$ 30,438 nT the dipole magnetic field intensity on the Earth surface at $r_0 = 1 R_E$); (ii) the contribution from a uniform field compression by the Chapman-Ferraro boundary currents; and (iii) a day-night asymmetry caused by the cross-tail current. The second and third terms are small compared to the dipole term. To first order, the coefficients b_1 and b_2 depend on the stand-off distance to the subsolar point of the magnetopause; the following relations represent this relationship reasonably well:

$$b_1 = 25\left(\frac{10}{R_s}\right)^3 \text{ nT}$$

$$b_2 = 2.1\left(\frac{10}{R_s}\right)^4 \text{ nT} \tag{1.50}$$

Notice the strong dependence with the stand-off distance R_s ($R_s = 10R_E$ corresponds to the normal state of the magnetosphere). The equation of the drift trajectory $B_0(r_0, \phi_0) =$ const. generated by a particle injected at point r_{0i}, ϕ_{0i} with a 90° pitch angle is, to first order:

$$r_0(\phi_0) = r_{0i} - \frac{R_E}{3}\frac{b_2}{B_E}\left(\frac{r_{0i}}{R_E}\right)^5 (\cos\phi_0 - \cos\phi_{0i}) \tag{1.51}$$

Note that only the day-night asymmetry coefficient b_2 appears. This equation represents eccentric circles with closest approach to the earth on the night side (as in Fig. 1.19); their eccentric displacement increases very rapidly with radial distance r_{0i}. Close to Earth higher order multipoles from the internal geomagnetic field must be taken into account for a more realistic description (Sect. 3.4).

Evaluating the drift velocity (1.46) of a particle along this $B_0 =$ const. contour, we find

$$V_D(\phi_0) = \frac{3mv^2}{2qR_E B_0}\left(\frac{B_0}{B_E}\right)^{1/3}\left(1 - \frac{4}{3}\frac{b_1}{B_0}\right) + \frac{5mv^2}{2qR_E}\frac{b_2}{B_0^2}\cos\phi_0 \tag{1.52}$$

This function passes through a maximum at midnight and a minimum at noon. Trapped equatorial particles thus indeed spend more time on the dayside than on the nightside of the magnetosphere, as we have anticipated above.

The drift period τ_d is given to first order by

$$\tau_d = \int_0^{2\pi} \frac{r_0 d\phi_0}{V_D} = \frac{4\pi q R_E^2}{3mv^2} B_0 \left(\frac{B_E}{B_0}\right)^{2/3}\left(1 + \frac{5}{3}\frac{b_1}{B_0}\right) \tag{1.53}$$

In this case only the compression coefficient b_1 intervenes. For particles of the same energy, τ_d increases as B_0 decreases (as the particle's drift contour radius increases), passes through a maximum for $B_0 \cong (20/3)b_1$ ($\sim$160 nT, or a radial distance of approximately 5.8 R_E), and then decreases again. The approximations used to derive these expressions start breaking down at these radial distances (for a more realistic field model the actual position of the contour of maximum drift period is slightly larger than the figure quoted.)

Our next examples involve the drift motion of equatorial particles subject to an *electric field* $\boldsymbol{E}_0 = -\nabla V$ parallel to the minimum-B surface (please *do not confuse* the electric scalar potential V with the drift velocity vector $\boldsymbol{V}_D$!). Each particle will be subjected to a drift that is the vector sum of (1.46) and (1.34):

$$\boldsymbol{V}_D = -M\nabla_\perp B_0 \times \frac{\boldsymbol{B}_0}{qB_0^2} - \nabla_\perp V \times \frac{\boldsymbol{B}_0}{B_0^2} = -\nabla_\perp\left(\frac{M}{q}B_0 + V\right) \times \frac{\boldsymbol{B}_0}{B_0^2} \tag{1.54}$$

For equatorial particles, this means that they will drift along curves of constant $\Psi = (M/q)B_0 + V$, which plays the role of an "extended" potential (note that $q\Psi = \mathfrak{E}$, total energy of the particle). The functions $B_0(r_0, \phi_0)$ and $V(r_0, \phi_0)$ can be determined using analytical or numerical models of the magnetic and electric fields. If the initial position of the particle's guiding center is r_{0i}, ϕ_{0i}, the drift trajectory $r_0 = r_0(\phi_0)$ is found by solving $\Psi = (M/q)B_0(r_0, \phi_0) + V(r_0, \phi_0) = (M/q)B_{0i} + Vi = \text{const.}$, where the magnetic moment $M = T_0/B_0 = T_{0i}/B_{0i}$ (non-relativistic case) serves to determine the particle's kinetic energy along its path. To calculate the time Δt to reach a given point of its drift path (or to determine the drift period in a closed orbit), it is necessary to integrate the inverse of (1.54): $\Delta t = \int (r_0/V_D) d\phi_0$ (dl: element of arc of the drift path).

We shall discuss examples using simple analytical approximations for B_0 and W_0. We choose a pure dipole field $B = B_E(R_E/r_0)^3$, and an electric field consisting of two terms: (i) an ubiquitous *corotational* field arising from the rotation of the dipole (co-axial rotation in this simplified model), and (ii) a uniform dawn-dusk electric field (assumed to be driven by the solar wind flow). The corotational field is an induced field with the property, mentioned on page 19, that any near-zero energy particle (regardless of its mass and charge) will drift with the local rotational speed ($v_{corot} = \Omega_E r_0$) in the dipole field: $E_{corot} = v_{corot} B_0 = \Omega_E {R_E}^3 B_E r_0^{-2}$ ($\Omega_E = 7.272 \times 10^{-5}$ rad/s) is the angular rotational speed of the Earth). Despite being an induced field (of the type $-\partial \boldsymbol{A}/\partial t$; see Appendix A.1), in the domain of interest this expression can be written as the gradient of a potential $V_{corot} = -\Omega_E {R_E}^3 B_E / r_0$. Concerning the uniform dawn-dusk electric field $\boldsymbol{E}_{dd}$, its potential is $V_{dd} = E_{dd} r_0 \sin\phi_0$ (positive, to have $\boldsymbol{E}$ pointing in the $-y$ direction). The final expression of Ψ is then

$$\Psi = (M/q)B_0 + V = (M/q)B_E(R_E/r_0)^3 - \Omega_E {R_E}^3 B_E/r_0 + E_{dd} r_0 \sin\phi_0 \quad (1.55)$$

Taking into account that $\nabla_\perp = \nabla_{\perp r} + \nabla_{\perp\phi}$ and that there is a vector product operation in (1.54), the corresponding drift velocity (1.54), in its polar components on the equatorial plane, is given by:

$$\begin{aligned} V_{Dr} &= -\frac{E_{dd}}{B_E {R_E}^3} r_0^3 \cos\phi_0 \\ V_{D\phi} &= -\frac{3M}{q\, r_0} + \Omega_E\, r_0 + \frac{E_{dd}}{B_E {R_E}^3} r_0^3 \sin\phi_0 \end{aligned} \quad (1.56)$$

Let us examine some characteristics of the equatorial particle drift paths $r_0 = r_0(\phi)$, solution of $\Psi = \Psi_i = \text{const.}$, where Ψ_i is defined by the initial values r_{0i}, ϕ_{0i} and $M = T_i/B_i$. We begin with analyzing the *electric* equipotentials, i.e., the drift trajectories of $M \to 0$ equatorial particles. Their equation will be, according to (1.55),

$$V = -\Omega_E {R_E}^3 B_E/r_0 + E_{dd} r_0 \sin\phi_0 = \text{const.} \quad (1.57)$$

Near the earth, the (negative) corotation potential will prevail ($V < 0$), and the equipotentials will approximate circles around the Earth; far away, e.g., toward

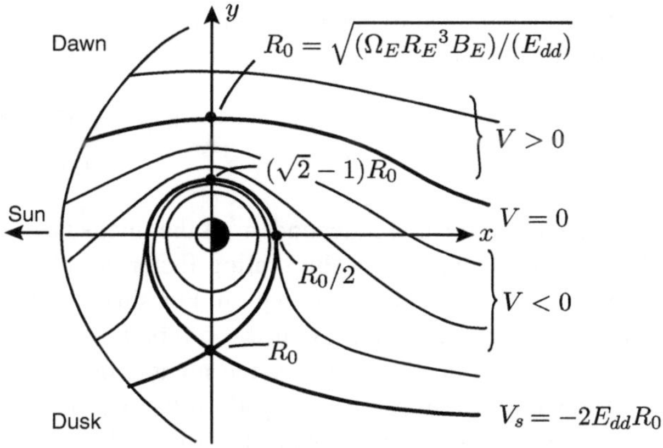

Fig. 1.20 Sketch of electric equipotentials on the magnetospheric equator for a magnetic dipole and a corotational plus dawn-dusk electric field. These equipotentials also represent the drift trajectories of near-zero energy equatorial particles (sunward convection and corotation)

the magnetotail, the equipotentials will approximate straight lines, representing a general sunward drift. The equation for the *zero* equipotential $V_0 = 0$ is $r_0{}^2 = (\Omega_E R_E{}^3 B_E)/(E_{dd} \sin\phi_0)$. Note that it is valid only for $\sin\phi_0 > 0$, i.e., it lies entirely in the dawn quadrants (see Fig. 1.20). Its intersection R_0 with the dawn meridian (y-axis) is located at

$$R_0 = \sqrt{(\Omega_E R_E{}^3 B_E)/(E_{dd})} \tag{1.58}$$

from the center of the Earth. This value will come handy as a geometric scaling parameter. The larger the electric field, the closer to the Earth the zero potential curve will come (the smaller the corotational region in the figure).

The solution of the quadratic equation (1.57) is

$$r_0(\phi_0) = \frac{V}{2E_{dd} \sin\phi_0}\left(1 \pm \sqrt{1 + \frac{4E_{dd}\Omega_E R_E{}^3 B_E}{V^2} \sin\phi_0}\right) \tag{1.59}$$

The presence of the function $\sin\phi_0$ in the square root indicates that the characteristics of the equipotentials in both dawn quadrants ($\sin\phi_0 > 0$) will in general be different from those in the dusk quadrants. Figure 1.20 sketches the different types of equipotentials for the general case. Since the value of r_0 must come out positive, in the dawn side there can only be one solution (the positive sign of the square root); in the dusk side two different solutions are possible. Note first the existence of two topological regions, a *corotation* region with closed equipotential lines around the Earth, and the *convection* region with open equipotentials, where low energy particles flow from the tail toward the front of the magnetosphere. The potential V_S of the *separatrix* limiting both regions will be that for which the two solutions along the dusk meridian ($\sin\phi_0 = -1$) coincide, i.e., for which the square root in (1.59) is zero: $V_S = -2\sqrt{E_{dd}\Omega_E R_E{}^3 B_E} = -2E_{dd}R_0$ (only the *negative* value of the square root will lead to positive r_0 in (1.59)). The equation of the separatrix will be, taking into account (1.58):

$$r_S(\phi_0) = -R_0 \frac{1 \pm \sqrt{1 + \sin\phi_0}}{\sin\phi_0} \tag{1.60}$$

The geometric form is independent of the intensity of the dawn-dusk electric field; the latter only appears in the scaling factor R_0 (1.58). Again, the larger E_{dd} the smaller will be the corotation region. For $\sin\phi_0 > 0$ only the negative sign of the square root is acceptable; for $\sin\phi_0 < 0$ both signs are possible, and two solutions may exist. Observe in the figure the notable points, intersections of the separatrix with the y-axis ($\sin\phi_0 = \pm 1$) and on the x-axis (for the latter, consider that the limit of the geometric factor in (1.60) for $\sin\phi_0 \to 0$ is $1/2$).

Some words about the concept of field line motion (page 20) in relation to the equipotentials shown in Fig. 1.20. As stated above, these equipotentials are the trajectories of near-zero energy charged particles, moving with a velocity (1.54) (without the M-term). Would this then also be the velocity of the corresponding field lines? Would the entire magnetic field be convecting and corotating as prescribed by the motion of the field lines' equatorial points, up to their intersection with the Earth? First of all, we haven't said anything about the electric field off the minimum-B equator. If all field lines are *equipotentials* ($\boldsymbol{E}_{\parallel} \equiv 0$) such a picture might indeed be correct. Yet in the definition of field line velocity, did we not require that the electric field be an *induced* electric field (1.38)? Well, just as the induced rotational electric field (page 19) can be expressed as the gradient of a scalar in the region of interest, the dawn-dusk electric field $\boldsymbol{E}_{dd}$ was also expressed as a potential field in the region of interest, despite also being an induced field (the moving or changing source currents to be found in the solar wind). So indeed we could visualize the curves of Fig. 1.20 as the trajectories of the equatorial footprints of magnetospheric field lines in our model. Unfortunately, their intersection with the conducting ionosphere substantially complicates the picture. Despite these reservations, the field lines through the closed portion of the separatrix may be viewed as an approximation to the limit of corotating low energy plasma particles that form the plasmasphere. Such limiting surface is called the *plasmapause*, and Eq. (1.60) may be viewed as "the plasmapause equation"; its form and dependence of a dawn-dusk electric field indeed bear some observed geometric and dynamic characteristics of the plasmapause.

We now turn to particles of arbitrary energy. If, say, $T_i \gtrsim 100\,\text{keV}$, the electric field drift can be neglected and electrons and protons drift in opposite senses along constant B lines, as discussed in relation to Fig. 1.19. In the intermediate range of energies the drift motion is more complicated. Electrons in general behave rather "normally" because in general the electric drift and the gradient drift point roughly in the same direction. For protons, however, these drifts may point in opposite directions on some parts of the dusk side; depending on which is the greater, the proton drift will be eastward or westward. Figure 1.21a, b show drift trajectories calculated using (1.54) for 1 keV electrons and protons, respectively, injected on the disk

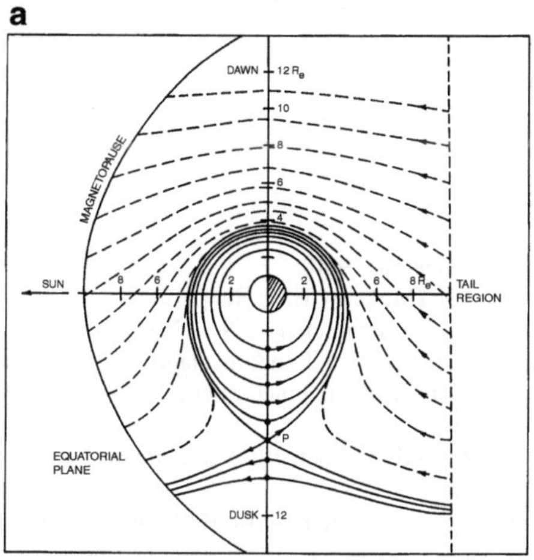

Fig. 1.21 (**a**) *Broken lines*: equipotentials for an electric field of potential $E_{dd} = 1.8\,\mathrm{kV}\ R_E^{-1}$. *Solid lines*: drift paths of electrons injected with 1 keV along the dusk meridian (*dots*). Notice corotation vs. convection. Equipotentials in the corotation region (not shown) are close, but not equal to the drift paths. (**b**) *Solid lines*: drift paths of protons injected with 1 keV along the dusk meridian (*dots*). Notice three types of paths: corotational, "vortices" not enclosing the earth, and sunward convection

meridian (solid curves), for a dawn-dusk electric field of 1.8 kV/R_E. Close to Earth, the electrons are stably trapped; beyond the stagnation point 1 keV electrons are quasi-trapped, being convected into the boundary. There is a reversal of drift sense at the stagnation point. Notice also how corotating electron drift paths approach the earth closer at dawn; their energy there can be up to ten times greater than at dusk. The behavior of protons (Fig. 1.21b), again injected with 1 keV at several positions along the dusk meridian, is more complex. Starting at 3, 4 or 5 R_E at dusk, the corotational field takes them eastwards around the earth in orbits similar to those of electrons; the energy-dependent gradient drift, directed opposite to the electric field drift, is negligible. Between 5 and 7 R_E on the dusk meridian, we have a zone in which protons get sufficiently accelerated in their eastward drift so that, eventually, the gradient drift takes over and turns them around westwards against the corotation drift, on the *same* evening side of the earth. After crossing the dusk meridian, these protons are decelerated and the electric field drift takes over again. We thus have closed drift paths which do not encircle the earth. Beyond 7 R_E on the dusk meridian, 1 keV protons are quasi-trapped, following a convection pattern toward the Sun.

Thus far we have assumed static conditions. Equation (1.54) can be integrated also for a slowly, adiabatically varying dawn-dusk field ($V = V(t)$). This can be used to determine the fate of low energy equatorial protons injected from the magnetospheric tail and convecting toward the Earth during conditions of high E_{dd}, and then captured around the Earth in corotating orbits when a decrease in E_{dd} sets in (having the effect of expanding the separatrix). This, in connection with more realistic field models and consideration of local plasma effects, may play a role in substorm ring current injection dynamics.

References

1. H. Goldstein, *Classical Mechanics* (Addison-Wesley, Boston, 1980)
2. B. Rossi, S. Olbert, *Introduction to the Physics of Space* (McGraw Hill, New York, 1970)
3. T.G. Northrop, *The Adiabatic Motion of Charged Particles* (Interscience, New York, 1963)
4. S. Eliezer, P. Norreys, J.T. Mendonça, K. Lancaster, Effects of Landau quantization on the equations of state in intense laser plasma interactions with strong magnetic fields. Phys. Plasmas **12**(5), 052115 (2005)
5. J.W. Belcher, S. Olbert, Field line motion in classical electromagnetism. Am. J. Phys. **71**, 220–228 (2003)
6. R.P. Feynman, R.B. Leighton, M. Sands, *Feynman Lectures on Physics*, vol. II (Addison-Wesley, Boston, 1963)
7. J.G. Roederer, Geomagnetic field distortions and their effects on radiation belt particles. Rev. Geophys. Space Phys. **10**, 599–630 (1972)
8. M. Schulz, L.J. Lanzerotti, *Particle Diffusion in the Radiation Belts* (Springer, Berlin, 1974)
9. D.H. Fairfield, Average magnetic field configuration of the outer magnetosphere. J. Geophys. Res. **73**, 7329–7338 (1968)

Chapter 2
Higher Order Drifts and the Parallel Equation of Motion

2.1 General Expression of the Drift Velocity and Higher Order Drifts

We are now in the position of defining a charged particle's guiding center system in a mathematically more rigorous, less qualitative, way. We shall use the three examples discussed in the preceding chapter (force drift, electric field drift, gradient-B drift) and realize that in each one, our implicit "recipe" was to look for a reference frame moving with velocity $\boldsymbol{V}$ in which, in a plane perpendicular to $\boldsymbol{B}$, the motion-induced electric field force $q\boldsymbol{V} \times \boldsymbol{B}$ cancels the phase-average of the resultant of all other forces acting on the particle as it makes one cyclotron turn. We can now generalize this rule in a more formal way.

First, we should make it clear that all that follows assumes the validity of the adiabatic conditions, which we write here in more general form in terms of the order of magnitude of characteristic space and time variations of any field quantity Q:

$$\rho_C \ll \frac{Q}{\nabla Q} \tag{2.1}$$

$$\tau_C \ll \frac{Q}{dQ/dt} \tag{2.2}$$

It is important to clearly understand the meaning and consequences of these conditions. Condition (2.1) tells us that the quantity Q should vary only very little (in relative magnitude and direction, if it is a vector) along the cyclotron trajectory of the particle (whose dimension depends on the particle's perpendicular velocity for a given magnetic field intensity (1.19)). Concerning the second condition, the time derivative is a total derivative, i.e., it represents change as seen by the particle. Therefore, (2.2) tells us that the quantity Q should change only very little (in relative magnitude or direction) during a cyclotron period (which in the non-relativistic domain does *not* depend on the particle's velocity).

J.G. Roederer and H. Zhang, *Dynamics of Magnetically Trapped Particles*, Astrophysics and Space Science Library 403, DOI 10.1007/978-3-642-41530-2_2,

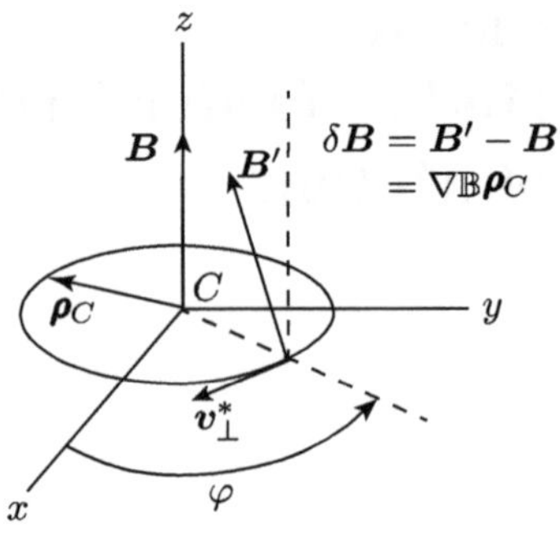

Fig. 2.1 Magnetic field configuration on a cyclotron orbit in the "natural" frame of reference

Second, we identify the type of forces that in most general terms may act on a charged particle at point $\boldsymbol{r}$ and time t in its cyclotron motion in the yet-to-be defined GCS. They are: (1) the external electric field force $q\boldsymbol{E}$ (electric field in the OFR); (2) a non-electromagnetic external force $\boldsymbol{F}$ (unimportant in magnetospheric physics); (3) the motion-induced electric field force $q\boldsymbol{V} \times \boldsymbol{B}$ ($\boldsymbol{V}$ is the yet-to-be determined velocity vector of the GCS); (4) the Lorentz force $q\boldsymbol{v}^* \times \boldsymbol{B}$ ($\boldsymbol{v}^*$ is the particle's velocity vector in the GCS); and as a new element, (5) the inertial force $-md\boldsymbol{V}/dt$ which appears whenever the GCS is an accelerated frame of reference.

Our formal definition can now be formulated in the following way: The guiding center system is a moving (usually non-inertial) frame of reference in which at any given time *the cyclotron phase average of all forces acting on the particle is zero.* Mathematically:

$$\left\langle q\boldsymbol{E} + \boldsymbol{F} + q\boldsymbol{V} \times \boldsymbol{B} + q\boldsymbol{v}^* \times \boldsymbol{B} - m\frac{d\boldsymbol{V}}{dt} \right\rangle \equiv 0 \tag{2.3}$$

Each term of this equation is a cyclotron phase average of the type

$$\langle P \rangle_\varphi = 1/2\pi \int_0^{2\pi} P(\varphi) d\varphi \tag{2.4}$$

Refer to Fig. 2.1, in which the origin is the instantaneous guiding center (position $\boldsymbol{r}_C$ in the OFR, (1.25)) and the axes represent the natural frame defined in Appendix A.1 (z-axis along $\boldsymbol{e} = \boldsymbol{B}/B$; x-axis along binormal $\boldsymbol{b}$; y-axis along normal $\boldsymbol{n}$). Since we are assuming the adiabatic condition (2.1) to apply, we can relate any vector $\boldsymbol{P}(\boldsymbol{r})$ on the particle's cyclotron orbit to its value at the guiding center via a first order Taylor expansion:

$$\boldsymbol{P}(\boldsymbol{r}) = \boldsymbol{P}(\boldsymbol{r}_C) + \delta\boldsymbol{P} = \boldsymbol{P}(\boldsymbol{r}_C) + \nabla \otimes \boldsymbol{P}|^T \boldsymbol{\rho}_C \tag{2.5}$$

where $\nabla \otimes \boldsymbol{P}|^T$ is the transposed tensor gradient of $\boldsymbol{P}$ (see (A.34) and footnote on page 170 of Appendix A.1) and $\boldsymbol{\rho}_C = \boldsymbol{\rho}_C(\boldsymbol{r}_C, \varphi)$ is given by (1.25). The last term has components $\nabla \otimes \boldsymbol{P}|^T \boldsymbol{\rho}_C|_i = \Sigma_k \partial P_i/\partial x_k \rho_{Ck}$.

The first and second term averages in (2.3) are just the electric field and the external force at the position of the guiding center $\boldsymbol{r}_C$, respectively ($\langle \delta \boldsymbol{E} \rangle = \langle \delta \boldsymbol{F} \rangle = 0$ because only $\sin\varphi$ and $\cos\varphi$ would appear in the integral (2.4)). The same applies to the third term ($\boldsymbol{V}$ is a quantity characterizing the whole GCS, not a vector field). In the fourth term, when we replace $\boldsymbol{B}(\boldsymbol{r})$ by an expansion like (2.5), we must take into account that for the vector velocity $\langle \boldsymbol{v}^* \rangle = 0$ in the GCS (1.7), so only the "little Lorentz force" $q\langle \boldsymbol{v}_\perp^* \times \boldsymbol{\delta B} \rangle$ would survive. The fifth term is an inertial force, which also should be considered a characteristic of the whole GCS.

The fourth term, $q\langle \boldsymbol{v}^* \times (\nabla \otimes \boldsymbol{B}\, \boldsymbol{\rho}_C)\rangle$ requires careful and detailed consideration. Consider a particle of mass m and electric charge q gyrating with a 90° pitch angle in a non-uniform magnetic field, as shown (for a positive charge) in Fig. 2.1. First we must find the expression of the tensor product. To that effect we note that by components,

$$\boldsymbol{v}^* = \left(\frac{q}{|q|}\, v_\perp^* \sin\varphi\,,\ -\frac{q}{|q|}\, v_\perp^* \cos\varphi\,,\ 0\right)$$

$$\boldsymbol{\rho}_C = (\rho_C\ \cos\varphi\,,\ \rho_C\ \sin\varphi\,,\ 0)$$

We inserted the ratio $q/|q|$ in order to put in evidence the effect of the sign of the electric charge.

The components of $\boldsymbol{\delta B} = \nabla \otimes \boldsymbol{B}|^T\ \boldsymbol{\rho}_C$ are (A.16):

$$\delta B_x = \frac{\partial B_x}{\partial x}\rho_C \cos\varphi + \frac{\partial B_x}{\partial y}\rho_C \sin\varphi$$

$$\delta B_y = \frac{\partial B_y}{\partial x}\rho_C \cos\varphi - \left(\frac{\partial B}{\partial s} + \frac{\partial B_x}{\partial x}\right)\rho_C \sin\varphi$$

$$\delta B_z = \nabla_x B\rho_C \cos\varphi + \nabla_y B\rho_C \sin\varphi$$

Remember that in the natural coordinate system $B_z = |B|$, $\partial B_z/\partial z = \partial B/\partial s$, and that $\nabla B = (\partial B/\partial x, \partial B/\partial y, \partial B/\partial s)$ and $\nabla \cdot \boldsymbol{B} = 0$ (Appendix A.1).

For $\langle \boldsymbol{v}^* \times \boldsymbol{\delta B} \rangle$ we need the component products $q v_i^* \delta B_k$ and average over one cyclotron turn. All terms containing $\cos\varphi \sin\varphi$ will average out to zero; in the remaining terms we will have integrals of the type $1/2\pi \int \sin^2\varphi d\varphi = 1/2\pi \int \cos^2\varphi d\varphi = 1/2$. The end results are:

$$q\langle \boldsymbol{v}^* \times \boldsymbol{\delta B} \rangle_x = \frac{1}{2} q\left(-\frac{q}{|q|}\right) v_\perp^* \rho_C \nabla_x B = -\frac{1}{2}|q| v_\perp^* \rho_C \nabla_x B$$

$$q\langle \boldsymbol{v}^* \times \boldsymbol{\delta B} \rangle_y = -\frac{1}{2}|q| v_\perp^* \rho_C \nabla_y B$$

$$q\langle \boldsymbol{v}^* \times \boldsymbol{\delta B} \rangle_z = -\frac{1}{2}|q| v_\perp^* \rho_C \frac{\partial B}{\partial s}$$

In vector form, we can write:

$$q\langle \boldsymbol{v}^* \times \boldsymbol{\delta B} \rangle = -\frac{m v_\perp{}^{*2}}{2B} \nabla B = -M \nabla B \tag{2.6}$$

The projection of this result perpendicular to $\boldsymbol{B}$ represents the *gradient-B force*,

$$\boldsymbol{f}_\perp = \boldsymbol{f}_g = -M \nabla_\perp B \tag{2.7}$$

which can be interpreted as the cause of the gradient-B drift ((1.45)—see also below). The parallel component of the average Lorentz force is what is called the *mirror force*

$$f_\parallel = -M \nabla_\parallel B = -M \frac{\partial B}{\partial s} \tag{2.8}$$

This force accelerates and decelerates particles spiralling along a field line into and away from decreasing or increasing B values, respectively. As we shall see later, it governs the bounce motion along a field line and is responsible for particle trapping in the geomagnetic field. Relations (2.7) and (2.8) provide further legitimacy to the model of a virtual guiding center particle and the concept of its magnetic moment: *any* magnetic moment $\boldsymbol{M}$ placed in a non-uniform magnetic field is subjected to a net force $\boldsymbol{f} = \nabla(\boldsymbol{M} \cdot \boldsymbol{B})$ which in our case is $= -M \nabla B$ because of (1.26) and the fact that M is an adiabatic invariant. One cautionary note is in order: we emphasized repeatedly that the velocity $v_\perp^*$ in the definition (1.26) of the magnetic moment is *not* that of the original particle in the OFR, but its transverse velocity in the GCS. When, in the GCS, this velocity is a function of the cyclotron phase the phase-average must be taken (see (1.41) and pertinent discussion). From now on, we shall drop the star supraindex to simplify the aspect of the equations, but we will remind the reader when necessary to distinguish between OFR and GCS variables.[1]

In summary, the condition (2.3) defining the guiding center system has now become

[1]It is advisable to revisit the above derivation process starting with Fig. 2.1 and relation (2.3). That process really developed in stages: what the figure intended to show implicitly was a "pre-GCS" in which the particle was circling free of *external* forces—i.e., a system which was moving with a transverse drift velocity, sum of $\boldsymbol{U}$ (1.34) and $\boldsymbol{V}_F$ (1.31) and in which the corresponding motion-induced field force balanced the external field forces. In such a system the particle gyrates with constant speed $v_\perp{}^*$ (example of Sect. 1.3). However, due to any inhomogeneity of the magnetic field, there was another resultant force, the term (2.6), which leads to an additional drift and the "final GCS". Now the particle's transverse velocity $v_\perp$ is no longer independent of φ (see expression (1.41)), but its cyclotron average is equal to the (constant) velocity in the "pre-GCS" (see expression (1.42)). It is precisely *this* average transverse velocity that enters in the definition of the magnetic moment (1.46). Confused again? Unfortunately, this detail is conceptually important, especially for the fundaments of plasma physics.

$$q\boldsymbol{E} + \boldsymbol{F} + q\boldsymbol{V} \times \boldsymbol{B} - M\nabla B - m\frac{d\boldsymbol{V}}{dt} \equiv 0 \tag{2.9}$$

The quantities involved are evaluated at the GC point, not at the original particle's actual position in its gyromotion. This equation will lead us to a dynamic equation of motion for the guiding center, to be discussed in the next section. At this time we focus on the fact that there is only one drift velocity that satisfies it (multiply the equation vectorially by $\boldsymbol{B}/qB^2$ and extract $\boldsymbol{V}_\perp$):

$$\boldsymbol{V}_D = \boldsymbol{V}_\perp = \left(q\boldsymbol{E} + \boldsymbol{F} - M\nabla B - m\frac{d\boldsymbol{V}}{dt}\right) \times \frac{\boldsymbol{B}}{qB^2} \tag{2.10}$$

This equation displays the four fundamental drifts of a guiding center particle: (1) the E-cross-B drift (1.34); (2) the force drift (1.31); (3) a gradient-B drift like (1.46); and (4) an inertial force drift.

For the parallel velocity of a guiding center particle, all we can do is repeat relation (1.4):

$$\boldsymbol{V}_\parallel = \langle \boldsymbol{v}_\parallel \rangle \cong \boldsymbol{v}_\parallel \tag{2.11}$$

Our last task is to further analyze the inertial term in (2.10). An apparent problem arises at once: this term contains the unknown vector $\boldsymbol{V}$ which we are trying to determine! But there are "unknown unknowns" and "known unknowns". This particularly applies to adiabatic theory because of its goal of providing useful approximations rather than impractical exactitude. If we divide the velocity into the two vectors $\boldsymbol{V} = \boldsymbol{V}_\parallel + \boldsymbol{V}_\perp$ we have:

$$\frac{d\boldsymbol{V}}{dt} = \frac{d(\boldsymbol{V}_\parallel + \boldsymbol{V}_D)}{dt} = \frac{d}{dt}(v_\parallel \boldsymbol{e} + \boldsymbol{V}_D) = \frac{dv_\parallel}{dt}\boldsymbol{e} + v_\parallel \frac{d\boldsymbol{e}}{dt} + \frac{d\boldsymbol{V}_D}{dt} \tag{2.12}$$

The second term calls our attention to the fact that the natural coordinate system (Appendix A.1) is a *local* frame of reference, that in an inhomogeneous magnetic field varies from point to point. In the above equation (and many future ones) the operator d/dt represents the total variation per unit time as seen by the particle, always consisting of two parts: (i) a variation in time at a fixed point in space, and (ii) a variation due to the displacement (which can be both perpendicular as well as parallel to $\boldsymbol{B}$) while the field is frozen in time: $d/dt = \partial/\partial t + \boldsymbol{V} \cdot \nabla$ (the latter operator being $\sum V_k \partial/\partial x_k$). We obtain the following:

$$\begin{aligned}
\frac{d\boldsymbol{V}}{dt} &= \frac{dv_\parallel}{dt}\boldsymbol{e} + v_\parallel\Big[\frac{\partial \boldsymbol{e}}{\partial t} + (\boldsymbol{V} \cdot \nabla)\boldsymbol{e}\Big] + \frac{d\boldsymbol{V}_D}{dt} \\
&= \frac{dv_\parallel}{dt}\boldsymbol{e} + v_\parallel\Big[\frac{\partial \boldsymbol{e}}{\partial t} + (\boldsymbol{v}_\parallel \cdot \nabla)\boldsymbol{e} + (\boldsymbol{V}_D \cdot \nabla)\boldsymbol{e}\Big] + \frac{d\boldsymbol{V}_D}{dt} \\
&= \frac{dv_\parallel}{dt}\boldsymbol{e} + {v_\parallel}^2 \frac{\partial \boldsymbol{e}}{\partial s} + v_\parallel(\boldsymbol{V}_D \cdot \nabla)\boldsymbol{e} + v_\parallel \frac{\partial \boldsymbol{e}}{\partial t} + \frac{d\boldsymbol{V}_D}{dt}
\end{aligned} \tag{2.13}$$

Carefully note that this is a purely kinematic/field-geometric equation—it has nothing to do with the dynamics of the particular particle involved, and might as well apply to little balls sliding without friction along bent and moving wires (the field lines)! Since $v_\| = ds/dt$, the first term is the "real" guiding center particle acceleration along the field line, measuring how its actual speed along the line varies; it is the one we are really interested in. The other terms are inertial accelerations due to the fact that a particle traveling along a bent and eventually moving field line experiences additional accelerations; they are important through the action of the inertial forces they represent. For instance, we recognize in the second term of the last equality a *centripetal acceleration* governed by the radius of curvature R_C of the field line in question (because according to (A.15) of Appendix A.1 $\partial \boldsymbol{e}/\partial s = -\boldsymbol{n}/R_C$); this indeed represents the fact that the guiding center follows the curved field line in its parallel motion. The third term is an inertial acceleration that appears if the guiding center has a *drift component along the normal* $\boldsymbol{n}$ of the field line; if δl is an element of the GC trajectory, this term can be written as $v_\| V_D \partial \boldsymbol{e}/\partial l$. The fourth term is an acceleration due to a *time-change of the direction* of the magnetic field (e.g., in rotating field lines). Finally, concerning the third and fifth term, in most of what follows we will replace $\boldsymbol{V}$ by the electric drift $\boldsymbol{U}$ (1.34) in order to avoid (to first order) the problem of the "unknown unknown" mentioned above. This replacement is more than just a convenience: it is fully justified as a legitimate approximation.

We now insert (2.13) (times the particle mass m) into the corresponding term on the right side of Eq. (2.10) to obtain the most complete expression of the drift velocity. With $\boldsymbol{U}$ and $\boldsymbol{V}_F$ given by (1.34) and (1.31), respectively, we write it in two equivalent forms:

$$\begin{aligned}\boldsymbol{V}_D &= \frac{\boldsymbol{e}}{qB} \times \left[-q\boldsymbol{E} - \boldsymbol{F} + M\nabla_\perp B + mv_\|^2 \frac{\partial \boldsymbol{e}}{\partial s} + mv_\| \frac{\partial \boldsymbol{e}}{\partial t} \right. \\ &\quad \left. + mv_\| (\boldsymbol{V}_D \cdot \nabla)\boldsymbol{e} + m\frac{d\boldsymbol{V}_D}{dt} \right] \\ &= \boldsymbol{U} + \boldsymbol{V}_F + \frac{m{v^2}_\perp}{2qB^3}\boldsymbol{B} \times \nabla_\perp B + \frac{m{v^2}_\|}{qB^2}\boldsymbol{B} \times \frac{\partial \boldsymbol{e}}{\partial s} \\ &\quad + \frac{m}{qB^2}\boldsymbol{B} \times \left[v_\| \frac{\partial \boldsymbol{e}}{\partial t} + v_\| (\boldsymbol{U} \cdot \nabla)\boldsymbol{e} + \frac{d\boldsymbol{U}}{dt} \right]\end{aligned} \tag{2.14}$$

The first equality (first two lines) expresses it all as a force drift made up of different contributions. The two lines in the second equality represent three different orders in the adiabatic approximation. Specifically, the first two lines show the zero-order E-cross-B and force drifts (1.34) and (1.31). The third line shows two first-order drifts:

$$\text{Gradient-B drift:} \quad \boldsymbol{V}_G = \frac{m{v^2}_\perp}{2qB^3}\boldsymbol{B} \times \nabla_\perp B = -\frac{\boldsymbol{M} \times \nabla_\perp B}{qB} \tag{2.15}$$

$$\text{Curvature drift:} \quad \boldsymbol{V}_C = \frac{m{v^2}_{\|}}{qB^2}\boldsymbol{B} \times \frac{\partial \boldsymbol{e}}{\partial s} \tag{2.16}$$

According to (A.22) in Appendix A.1, *in absence of any currents* $B\partial \boldsymbol{e}/\partial s = \nabla_\perp B$, thus under this special condition, we can combine the last two equations into one:

$$\text{Gradient-curvature drift:} \quad \boldsymbol{V}_{GC} = \frac{1}{2}\frac{m}{qB^3}({v^2}_\perp + 2{v^2}_{\|})\boldsymbol{B} \times \nabla_\perp B \tag{2.17}$$

The fourth line in the second equality of (2.14) contains second and higher order drifts; in the latter we have replaced $\boldsymbol{V}_D$ with its own first approximation, $\boldsymbol{U}$ (for $\boldsymbol{F} = 0$). The first term represents the effect of time-dependence of the direction of the magnetic field; the second represents the effect of a spatial variation of the $\boldsymbol{e}$ direction.

The last term is of importance in some electric field situations, in particular, for a stationary magnetic field but varying electric field. Taking into account (1.34), we obtain the so-called

$$\text{Polarization drift:} \quad \boldsymbol{V}_P = \frac{m}{qB^2}\frac{\partial \boldsymbol{E}_\perp}{\partial t} \tag{2.18}$$

Finally, a few nitpicking remarks about adiabatic drifts. What exactly are the particle velocities that appear in the general expression (2.14)? We strongly recommend that the reader review again the statements in the footnote on page 38 and all its implications. Another important point is the following. Carefully observe how the various terms in Eqs. (2.14)–(2.18) depend on the velocities $v_\perp$ and $v_\|$ of the original (real!) particle. Zero order drifts do not depend on them at all, only on the local $\boldsymbol{B}$, $\boldsymbol{E}$ and force fields. The gradient-B drift depends on M (see its second expression in (2.15)), which is an adiabatic constant of motion, i.e., depends only on the *initial* transverse velocity of the particle. The curvature drift *does* depend on its instantaneous $v_\|$. What enters in the remaining higher order drifts is the full drift velocity itself—but as stated above, for a second order approximation, the known drift U suffices.

2.2 Motion Along the Field Line and the Energy Equation

Before we discuss the dynamics of guiding center particles in their field-aligned motion, we should answer a fundamental question: *Why* does the guiding center particle actually follow a curved field line? Usually, this is a fact taken for granted, based on the appearance of a centripetal acceleration term ${v_\|}^2(\partial \boldsymbol{e}/\partial s)$ in (2.13) and confirmed by the beautiful pictures of glowing plasma-filled coronal loops. However, as stated before, this equation is a purely kinematic and field-geometric one, having to do with the way we represent our guiding center velocity vector in the

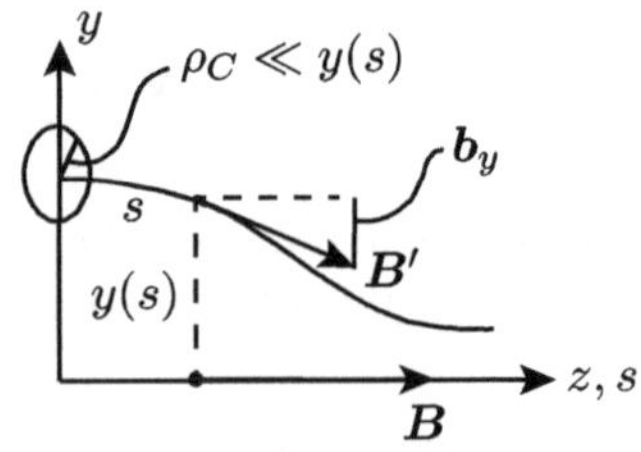

Fig. 2.2 Motion of the GC along a curved field line: geometric parameters

natural frame of reference (2.12). A more legitimate proof is the following. Consider a magnetic field configuration as shown in Fig. 2.2, with a central rectilinear field line along axis z and a magnetic intensity that increases toward positive z. A probe particle is placed on a near-by field line with small perpendicular velocity $\boldsymbol{v}_\perp$, therefore also with a small gyroradius $\rho_C \ll y(s)$. Assuming cylindrical symmetry around the z axis, the differential equation of the particle's field line (see Appendix A.1, footnote on page 164) can be easily derived from the flux tube property $By^2 = \text{const.}$ along z:

$$\frac{dy}{ds} = -\frac{1}{2B}\frac{\partial B}{\partial s}y$$

Now we determine the trajectory of the GC particle under the drift imposed by the "little Lorentz force" $q(\boldsymbol{v}_\parallel \times \boldsymbol{b}_y)$ directed along x (into the paper):

$$\boldsymbol{V}_D = q(\boldsymbol{v}_\parallel \times \boldsymbol{b}_y) \times \frac{\boldsymbol{B}}{qB^2} \quad \text{Therefore} \quad V_{Dy} = \frac{dy}{dt} = \frac{v_\parallel}{B}b_y$$

But $b_y = (\partial B_y/\partial y)y = -1/2(\partial B/\partial s)y$, because of $\nabla \cdot \boldsymbol{B} = 0$ and the symmetry around the z axis. Thus, the equation of the GC trajectory will be:

$$\frac{dy}{dt} = \frac{dy}{ds}\frac{ds}{dt} = \frac{dy}{ds}v_\parallel = -v_\parallel \frac{1}{2B}\frac{\partial B}{\partial s}y \tag{2.19}$$

Cancelling the particle variable $v_\parallel$ on both sides of the last equality leaves us with the differential equation of the particle's magnetic field line! (Footnote, page 164). In summary, under adiabatic conditions the "little Lorentz force" caused by any small non-uniformity $\boldsymbol{b}$ of the magnetic field acting on the parallel motion of a charged particle is always such as to impart an acceleration $\perp \boldsymbol{B}$ that makes the guiding center follow the curved field line.

There is a limit, though, to the property of "following the field line", imposed by adiabatic condition (2.2). If $\sigma_C = v_\parallel \tau_C$ is the displacement of the GC along a field line during one cyclotron turn, this condition requires that the magnetic field, including its direction $\boldsymbol{e}$, change very little along σ_C. This implies $\sigma_C(\partial e/\partial s) \ll 1$ or $\sigma_C \ll R_{fl}$, where R_{fl} is the field line's radius of curvature. For small pitch angles ($v_\parallel \simeq v$) the particle velocity must then be $v \ll R_{fl}/\tau_C$. Note that rather than a restriction on pitch angles, this is a restriction on velocity or energy: particles with small pitch angles and too high energy will "fly off" tangentially from their home field line (happens to trapped particles on sharply bent field lines in the neutral sheet)! This is equivalent to what happens to energetic particles with pitch angles $\approx 90°$: when condition (2.1) is violated, they will "fly off" perpendicularly to $\boldsymbol{B}$ (happens to trapped particles in the vicinity of the dayside boundary)!

We now go back to the guiding center system definition (2.9) and extract the time derivative of the GC velocity $\boldsymbol{V}$ in its parallel projection:

$$m\frac{d\boldsymbol{V}}{dt}\cdot\boldsymbol{e} = a_{\parallel} = qE_{\parallel} + F_{\parallel} - M\frac{\partial B}{\partial s} \tag{2.20}$$

Then we turn to (2.13) and consider its parallel projection

$$\left.\frac{dV}{dt}\right|_{\parallel} = (d\boldsymbol{V}/dt)\cdot\boldsymbol{e} = a_{\parallel} = \frac{dv_{\parallel}}{dt} + \frac{d\boldsymbol{V}_D}{dt}\cdot\boldsymbol{e} = \frac{dv_{\parallel}}{dt} - \boldsymbol{V}_D\cdot\frac{d\boldsymbol{e}}{dt}$$

the latter equality because of $\boldsymbol{V}_D\cdot\boldsymbol{e} = 0$. But $d/dt = \partial/\partial t + (\boldsymbol{V}\cdot\nabla)$ which leaves the above equation as

$$\left.\frac{dV}{dt}\right|_{\parallel} = a_{\parallel} = \frac{dv_{\parallel}}{dt} - v_{\parallel}\boldsymbol{V}_D\cdot\frac{\partial\boldsymbol{e}}{\partial s} - \boldsymbol{V}_D\cdot\frac{\partial\boldsymbol{e}}{\partial t} - \boldsymbol{V}_D\cdot(\boldsymbol{V}_D\cdot\nabla)\boldsymbol{e} \tag{2.21}$$

Note carefully that $a_{\parallel}$ is *not* d^2s/dt^2! Inserting (2.21) in (2.20), we finally obtain the *parallel equation of motion* of the guiding center:

$$\begin{aligned} m\frac{dv_{\parallel}}{dt} = m\frac{d^2s}{dt^2} &= qE_{\parallel} + F_{\parallel} - M\frac{\partial B}{\partial s} \\ &+mv_{\parallel}\boldsymbol{V}_D\cdot\frac{\partial\boldsymbol{e}}{\partial s} + m\boldsymbol{V}_D\cdot\frac{\partial\boldsymbol{e}}{\partial t} + m\boldsymbol{V}_D\cdot(\boldsymbol{V}_D\cdot\nabla)\boldsymbol{e} \end{aligned} \tag{2.22}$$

The first line includes zero- and first-order terms in which we recognize the external forces and the mirror force; the second line contains higher-order terms, corrections that are the result of field-geometric effects related to the drift velocity.

Let us give just two quick examples for the action of the second order 4th and 5th terms in (2.22) (quick, because they can be neglected in radiation belt physics), consider Figs. 2.3 and 2.4, respectively. The first one depicts a particle spiralling about the central line in a nearly uniform magnetic field in a long rotating solenoid. As viewed from the OFR (which is fixed to the rotating solenoidal field), the particle is subjected to an induced electric field $\boldsymbol{E}_i = -(\boldsymbol{\Omega}\times\boldsymbol{r})\times\boldsymbol{B}$ which in turn causes a drift $\boldsymbol{V}_D$, responsible for the particle corotating with the solenoid (see discussion on page 19). The term $m\boldsymbol{V}_D\cdot\partial\boldsymbol{e}/\partial t = m\Omega^2 r$ is the centrifugal force leading to a "sling shot effect" on the GC particle. The second example is a particle spiralling along a circular field line around a linear current I and subjected to a uniform electric field parallel to the current, out of the paper. Besides the gradient-B and curvature drifts

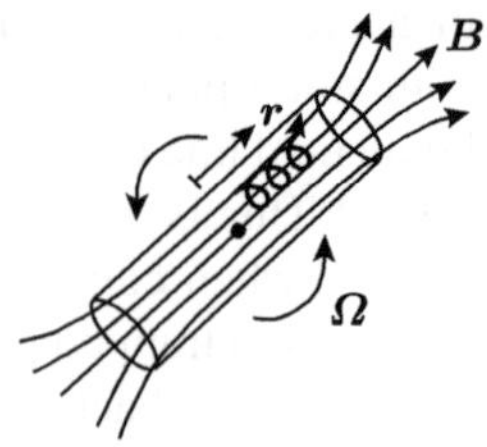

Fig. 2.3 Example of particle drifting in a rotating solenoid field

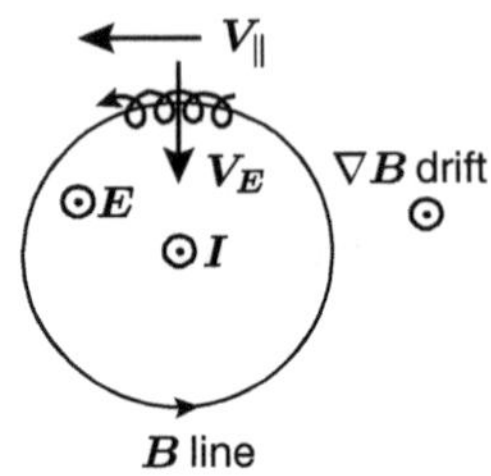

Fig. 2.4 Example of particle drifting in the field of a linear current I with an electric field along it

perpendicular to the paper, there will be an E-cross-B drift toward the center line, in the direction of $\partial e/\partial s$. So the 4th term will be $\neq 0$, representing an acceleration as the GC particle drifts into higher B-values.

In absence of any drift, or if all three higher order terms can be neglected, and if the external forces derive from a scalar potential *along* the field line W(s), the parallel equation of guiding center motion is reduced to

$$m\frac{dv_\parallel}{dt} = -\frac{\partial}{\partial s}\left[MB(s) + W(s)\right] \tag{2.23}$$

The guiding center moves along the field line as if it was subjected to a total scalar parallel potential $P_\parallel = MB(s) + W(s)$, and one can apply to that parallel motion all the familiar energy graphs and related methods of mass point mechanics (see next section).

Finally, there is one more important general equation to be derived. Consider the instantaneous kinetic energy of a gyrating particle in the OFR:

$$T = \frac{1}{2}mv_\perp^2 + \frac{1}{2}mv_\parallel^2$$

Taking into account relations (1.8), (1.26) and (2.11), we can write for the cyclotron average:

$$\langle T\rangle_C = MB + \frac{1}{2}mV_\parallel^2 + \frac{1}{2}mV_D^2 \tag{2.24}$$

In the guiding center particle model, the first term represents the GC particle's "internal energy" (cyclotron energy), the second term represents the kinetic energy of the GC particle in its motion along the field line, the third term the kinetic energy in its drift motion. The *energy equation* refers to the time rate of change of the GC particle's average kinetic energy, which we write:

$$\frac{d\langle T\rangle_C}{dt} = M\frac{dB}{dt} + mv_\parallel\frac{dv_\parallel}{dt} + m\boldsymbol{V}_D\cdot\frac{d\boldsymbol{V}_D}{dt} \tag{2.25}$$

Setting in the first term $d/dt = \partial/\partial t + v_\parallel\partial/\partial s + \boldsymbol{V}_D\cdot\nabla$ and replacing $dv_\parallel/dt$ with its expression in (2.22), we have:

$$\frac{d\langle T\rangle_C}{dt} = M\frac{\partial B}{\partial t} + V_\parallel(qE_\parallel + F_\parallel) + \boldsymbol{V}_D\cdot\left(M\nabla_\perp B + m{v_\parallel}^2\frac{\partial \boldsymbol{e}}{\partial s} + mv_\parallel\frac{d\boldsymbol{e}}{dt} + m\frac{d\boldsymbol{V}_D}{dt}\right)$$

The parenthesis appears as part of the expression of $\boldsymbol{V}_D$ in (2.14). Extracting $md\boldsymbol{V}_D/dt$ from (2.13) and then $d\boldsymbol{V}/dt$ from (2.9), the last term of the above equation turns out to be equal to $\boldsymbol{V}_D\cdot(q\boldsymbol{E}_\perp + \boldsymbol{F}_\perp)$, which leaves the energy equation as

$$\begin{aligned}\frac{d\langle T\rangle}{dt} &= M\frac{\partial B}{\partial t} + V_\parallel(qE_\parallel + F_\parallel) + \boldsymbol{V}_D\cdot(q\boldsymbol{E}_\perp + \boldsymbol{F}_\perp)\\ &= M\frac{\partial B}{\partial t} + \boldsymbol{V}\cdot(q\boldsymbol{E} + \boldsymbol{F})\end{aligned} \tag{2.26}$$

This result is easy to understand intuitively, and again confirms the physical adequacy and conceptual value of the guiding center particle model.[2] The first term is the power delivered by a changing magnetic field to the internal state of the virtual GC particle (the cyclotron motion of the real particle); note that it only includes the *local* time derivative of B. The agent responsible for the delivery or extraction of power is the induced electric field associated with the *locally* changing magnetic field $\boldsymbol{E}_{ind} = -\partial\boldsymbol{A}/\partial t$ (page 176 in Appendix A.1). This induced electric field acts on the real particle in its cyclotron motion, as sketched in Fig. 1.15.[3] Although trivial, we still must point out that changes in B as seen by the guiding center particle (i.e., convective changes in an inhomogeneous field $\boldsymbol{V}\cdot\nabla B$ or $v_\parallel\partial B/\partial s$) do not enter in the first term. For instance, a particle drifting in a static magnetic and electric field like in Fig. 1.21 does experience a varying B-field (in the GCS) because it is driven into it by the electric drift (otherwise the guiding center particle would follow a $B =$ const. contour, like in Fig. 1.19); whenever that electric field has a component

[2]For the relativistic version, use (1.29) as the relation between the relativistic magnetic moment and kinetic energy.

[3]When the local time derivative $\partial B/\partial t$ is entirely due to changes in the external current intensities, but not their configuration in space, one usually calls this process a *betatron acceleration*; if the current intensities are constant, but their position or distribution changes in space, one calls it a *Fermi acceleration*. This distinction is made mainly in astrophysics. However, the particle doesn't care about *what* causes the local field to change!

in the drift direction ($\boldsymbol{V}_D \cdot \boldsymbol{E} \neq 0$), it will do work on the particle. Another trivial but relevant remark: drifts due to the electric field do not play any role in the second term, because by definition (1.34) their contribution to the total drift $\boldsymbol{V}_D$ is *always perpendicular* to $\boldsymbol{E}$. This is particularly important to keep in mind when one deals with purely induced electric fields in absence of potential electric fields: we must include the induced electric field in the second term, but *only non-electric drifts* can lead to energy change (if they have a component along $\boldsymbol{E}_{ind}$). This sometimes is called drift-betatron, in distinction from the above gyro-betatron.

2.3 Particle Trapping and Parallel Electric Fields

In the previous section we have derived three fundamental and most general equations that under the adiabatic conditions (2.1) and (2.2) describe the dynamics of a virtual guiding center particle of given mass, charge, field-aligned velocity and magnetic moment. For the perpendicular motion, the *drift velocity* in its most general form is given by (2.14); we must emphasize again that this drift velocity does not appear as the result of the integration of a dynamic equation but, rather, is defined as the result of an averaging process, which then allows the replacement of the rapidly gyrating original particle by a virtual particle at the guiding center. For the parallel motion, we do have a real dynamic equation (2.22) determining the average acceleration of the particle along a field line. As a corollary, a third equation was derived, giving the average rate of change of the kinetic energy of the guiding center particle (2.26).

In this section we will focus on the parallel motion along a field line and, for that purpose, consider cases in which the particle drift is either a priori zero (say, for symmetry reasons), or negligible with respect to the parallel motion ($V_D \ll v_\parallel$). For instance, consider the field geometry of a "mirror machine", shown at right in Fig. 2.5, with the guiding center particle along the z-axis field line. But here comes a disappointment: instead of using the laboriously derived dynamic parallel equation (2.22), we turn to the following relations based on two simple conservation principles:

$$\text{Conservation of magnetic moment } M\text{:} \quad v_\perp^2(s) = \frac{2M}{m} B(s) \tag{2.27}$$

$$\text{Conservation of total energy } \mathfrak{E}\text{:} \quad v_\parallel^2(s) + v_\perp^2(s) = \frac{2}{m}\left[\mathfrak{E} - W(s)\right] \tag{2.28}$$

$\mathfrak{E}$ is the total mechanical energy, $W(s)$ is the potential energy. The initial conditions determine the constants: $M = T_i/B_i$ and $\mathfrak{E} = T_i + W_i$, in which the subindex i denotes value of the variable at the initial field line point s_i.

Of more practical significance are the equivalent equations in the variables $v_\perp$, $v_\parallel$ and pitch angle α (1.14), and their respective initial values (see Fig. 2.6):

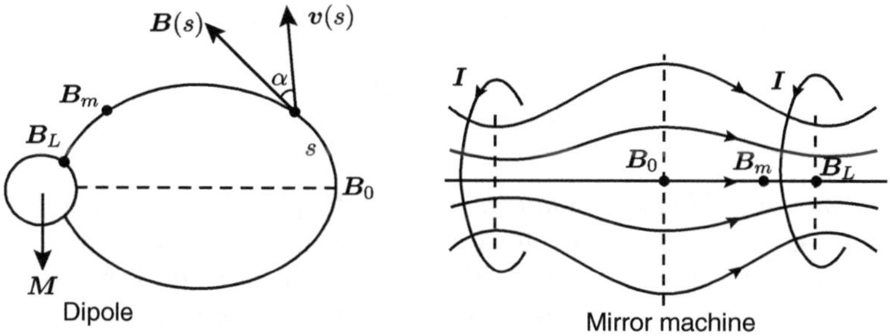

Fig. 2.5 Two fundamental types of trapping magnetic field geometries

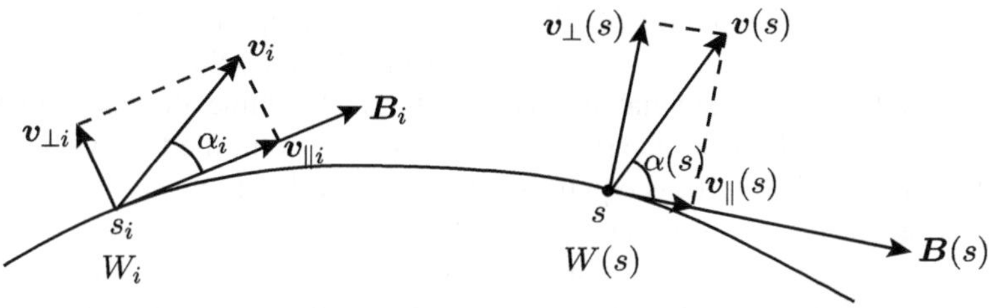

Fig. 2.6 Change of $\boldsymbol{v}_\perp$ and $\boldsymbol{v}_\parallel$ vectors along a magnetic field line

$$v_\perp^2(s) = v_i^2 \frac{\sin^2\alpha_i}{B_i} B(s) \tag{2.29}$$

$$v_\parallel^2(s) = v_i^2\big(1 - \frac{\sin^2\alpha_i}{B_i} B(s)\big) - \frac{2}{m}\big(W(s) - W_i\big) \tag{2.30}$$

$$\sin^2\alpha(s) = \frac{\sin^2\alpha_i}{B_i} B(s)\left[1 - \frac{W(s) - W_i}{T_i}\right]^{-1} \tag{2.31}$$

Note that the function $v_\perp(s)$ (and therefore the perpendicular kinetic energy $T_\perp$) is governed *only* by the local magnetic field intensity $B(s)$ (and the initial conditions)—it cannot be influenced by external forces! To change it, the conservation of magnetic moment M has to be violated, e.g., by scattering or resonance with cyclotron-resonant waves.

Our first example will be one in which field-aligned forces are absent, i.e., when magnetic field lines are equipotentials: $W(s) = W_i = \text{const.}$, $T = \text{const.}$ and $v = \sqrt{v_\perp^2 + v_\parallel^2} = \text{const.}$, and with no time variations of $\boldsymbol{B}$. Equation (2.31) is now

$$\sin^2\alpha(s) = \frac{\sin^2\alpha_i}{B_i} B(s) \tag{2.32}$$

The pitch angle $\alpha(s)$ increases toward higher $B(s)$ values and becomes $\pi/2$ when

$$B_m(s_m) = \frac{B_i}{\sin^2 \alpha_i} \tag{2.33}$$

B_m is the *mirror field* and s_m a *mirror point* of the particle on the field line. Field line points on which $B > B_m$ are off limits: the guiding center particle stops instantaneously at s_m ($v_\parallel = 0$ there, with all the motion in cyclotron mode: $v_\perp = v$), and then reverses its sense. To understand why the GC particle doesn't just slow down and stop dead, we must turn to the parallel equation (2.22), which in this particular case is simply $m\, dv_\parallel/dt = -M\, \partial B/\partial s$: it is the mirror force (2.8) that turns the particle around in its parallel motion. If we inject a particle with a 90° initial pitch angle at any point s of a field line, this initial point will also be a mirror point and the mirror force will start accelerating the particle along the field line toward lower B-values. Note that for an equipotential field line, the mirror point field intensity B_m is an adiabatic invariant because at such point the transverse velocity in the definition of M (1.26) is equal to the constant velocity v.

In general, in field configurations like the dipole field or a mirror machine in Fig. 2.5, there are two mirror points on either side of any initial point s_i: as a particle reverses its parallel motion at a mirror point, it will move toward lower B values, which will pass through a minimum (see below) and increase again; as a consequence, another mirror point may eventually be reached. The end result of all this is that a guiding center particle is *trapped*, bouncing back-and-forth between two *conjugate* mirror points. The *bounce period* of a particle trapped between two mirror points s_m and s'_m on an equipotential field line will be, according to (2.30),

$$\tau_b = 2\int_{s'_m}^{s_m} \frac{ds}{v_\parallel(s)} = \frac{2}{v}\int_{s'_m}^{s_m} \frac{ds}{\left[1 - B(s)/B_m\right]^{\frac{1}{2}}} \tag{2.34}$$

The integral is extended along the field line between mirror points located at s'_m and s_m, where the field intensity B_m is given by (2.33). The integral

$$S_b = \frac{1}{2} v\, \tau_b = \int_{s'_m}^{s_m} \frac{ds}{\left[1 - B(s)/B_m\right]^{\frac{1}{2}}} \tag{2.35}$$

is the *half-bounce path*, i.e., the rectified path of the original cycling particle between one mirror point and its conjugate. Note that it is a purely field-geometric quantity, independent of the particle in question. We can think of it as a function of space, a scalar field $S_b(s_m)$ representing the rectified inter-mirror-point path of a trapped particle mirroring at *that* point. Unfortunately, even for simple field geometries like a dipole field, the integral in (2.34) and (2.35) cannot be expressed in analytical, closed form and in practice must be calculated numerically. In Sect. 3.1 we will use the following relationship which will help overcome the numerically annoying fact that the integrand in the above relations has an integrable singularity at the mirror points:

$$S_b = I + 2B_m \frac{\partial I}{\partial B_m}\bigg|_{fl} \tag{2.36}$$

where

$$I = \int_{s'_m}^{s_m} \left[1 - B(s)/B_m\right]^{\frac{1}{2}} ds \tag{2.37}$$

This function, whose integrand is the inverse of the integrand in (2.35) is easier to calculate numerically. Moreover, it is directly related to the second adiabatic invariant, as we shall show in the next section. The derivative with respect to B_m in (2.36) is to be taken on I as a function of the mirror point field intensity B_m *on the given field line*. Notice that not only the integrands of (2.35) and (2.37) are functions of B_m but also their integration limits, and that $S_b > I$ always.

Another notable point on a field line is where B has a local minimum: the pitch angle there will be minimum, too, whereas $v_\parallel$ will be maximum. In a dipole or dipole-like field, the minimum-B point is called the field-line's *equatorial point* (whether or not its geographic latitude actually is zero.) Note that in absence of parallel forces *all* trapped particles on a field line must transit through it. For this reason it is preferentially chosen as a fundamental reference point and origin for the curvilinear coordinate s (with positive values increasing toward the North in the geomagnetic case). Instead of α we shall work with the pitch angle variable $\mu = \cos\alpha$, commonly used in magnetospheric physics for reasons that will become apparent later. Equations (2.32) and (2.33) referred to the minimum-B point of a field line will now be:

$$\mu^2(s) = 1 - (1 - \mu_0^2)\frac{B(s)}{B_0} \quad \text{and} \quad B_m = \frac{B_0}{1 - \mu_0^2} \tag{2.38}$$

By definition, at a minimum-B point $\partial B/\partial s = 0$; in its neighborhood the function $B(s)$ can be approximated as $B(s) \simeq B_0 + 1/2as^2$, where $a = \partial^2 B/\partial s^2$ at s_0. Near the minimum-B point the parallel equation of motion (2.22) now becomes

$$m\frac{dv_\parallel}{dt} = -M a\, s \tag{2.39}$$

which indeed looks like that of an harmonic oscillator with a constant $k = Ma$. If we inject a GC particle with a 90° pitch angle at the minimum-B point, it will stay there in an equilibrium position (just cyclotron circling—remember that for the time being we are ignoring any drifts, but if we do take them into account, the GC particle will stay on a minimum-B surface even if the latter is slightly warped). If the pitch angle is slightly less than 90°, the particle will bounce in harmonic field-aligned motion between very close-by mirror points[4] with a bounce period derived from (2.39):

$$\tau_b = 2\pi \sqrt{\frac{m}{Ma}} \approx \frac{2\pi\sqrt{2}}{v}\sqrt{\frac{B}{a}} \tag{2.40}$$

The second near-equality is justified because $v_\perp \approx v$ in the expression of the magnetic moment M. Note that this period (or the bounce angular frequency $\omega_b = 2\pi/\tau_b$) to first order only depends on the local magnetic field geometry and the particle's (mostly transverse) velocity. Under the adiabatic condition (2.2), the bounce period is always much larger than the cyclotron period (1.20), $\tau_b \gg \tau_C$. The half-bounce path $S_b = 1/2\, v\tau_b$ is independent of the particle and it, too, is much larger than the Larmor radius under adiabatic condition (2.1). Note the apparently curious fact that even a 90° equatorial particle does have a finite bounce period and a half-bounce path—but that's the same thing as a mechanical oscillator in equilibrium position having a non-zero fundamental frequency despite being at rest! It can be shown that for near-equatorial particles, $a = B/{R_C}^2$, where R_C is the field line curvature at the equatorial point. The bounce path (2.35) and bounce period are then:

$$S_b = \pi\sqrt{2}R_C \quad \text{or} \quad \tau_b = (2\pi\sqrt{2}/v)R_C \tag{2.41}$$

Finally, there will always be points on a field line beyond which a particle is lost (intersection with the ionosphere in the geomagnetic field, maximum-B point in the coils of a mirror machine, Fig. 2.5). This leads to the concept of *loss cone* α_L at point s, which in absence of field-aligned electric field and forces has an aperture

$$\sin^2\alpha_L(s) = \frac{B(s)}{B_L} \tag{2.42}$$

where B_L is the magnetic field intensity at point s_L where the field line intersects the ionosphere (or passes through the coil in a mirror machine). At any point of a field line in the examples of Fig. 2.5 there will be always two loss cones, one for direction $+s$ and one for $-s$. If there is complete hemispheric symmetry of the field intersections with the ionosphere, or for a mirror machine along the z-axis, the two loss cones will be equal.

Our second example will consist of a geomagnetic dipole-like field line with a *parallel electrostatic potential* $V(s)$ (again, please do not confuse this symbol with drift velocity!). Now we must use the more general Eqs. (2.29)–(2.31). The right-hand side of the third equation consists of two factors, $(\sin^2\alpha_i/B_i)B(s)$,

[4] This justifies the entire discussion in Sect. 1.6 of equatorial particles: even if their pitch angles deviate a bit from 90°, during their drift they will always be tied to an equilibrium position on the minimum-B surface.

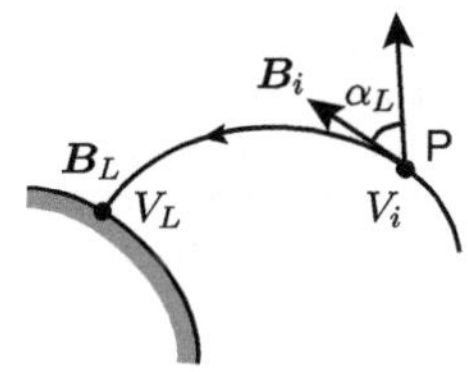

Fig. 2.7 Field parameters and particle loss cone in the northern hemisphere of a terrestrial field line

which is the value of $\sin^2\alpha(s)$ in absence of parallel forces, and an energy-dependent correction function $[1 - q(V(s) - V_i)/T_i]^{-1}$. Under these circumstances the expressions of the bounce period (2.34) and half-bounce path (2.35) have to be appropriately modified and the position of the mirror points s_m will be energy-dependent, solution(s) of the following equation directly derived from (2.31):

$$\frac{B(s_m)}{B_m^*} = 1 - q\frac{(V(s_m) - V_i)}{T_i} \tag{2.43}$$

In this expression $B_m^* = B_i/(1 - \mu_i^2)$, the mirror point field intensity in absence of field-aligned forces.

Let us examine the case in detail. The result of Eq. (2.31) (which always should be compromised between 0 and +1) may "go wrong" in two ways, representing an off-limits place for the particle. First, it could be > 1 because of too large values of B(s) (as it happened in our previous discussion of mirror points regarding (2.33)), or because of the energy-dependent correction function. Second, it could turn out < 0 because of this correction function between brackets. We shall discuss these situations in more detail.

First, to the mirror points. Refer to Fig. 2.7; at point P (arc-length s_i from the equatorial point) a particle is injected with kinetic energy T_i and pitch angle α_i traveling toward the Earth. L is the intersection with a loss region where the potential is V_L. First assume that $qV_L > qV_i$, which corresponds to a force directed toward the equator, away from L (for electrons, it would be an electric field directed *toward* the Earth). If $T_i < q(V_L - V_i)$, according to (2.31) the value of $\sin^2(\alpha_L)$ would be negative, i.e. the point s_L would be inaccessible to the particle. This means that *all* particles through P, even those with a 0° pitch angle, would mirror before reaching the loss region—turned around by the *combined* action of the electric force and the mirror force. On the other hand, if $qV_L < qV_i$ (parallel electric force directed toward the loss point L), $\sin^2(\alpha_L) > B/B_L$ and the loss cone at P will be bigger than in absence of the electric field. Thus, in general terms, decelerating potentials will narrow the loss cone (maybe even eliminate it), while accelerating potentials will widen it. This has important consequences for the action of an auroral mechanism which in general involves the generation of a field-aligned electric field directed from the ionosphere toward the equatorial point of a field line ($qV_L > qV(s)$), in both hemispheric branches of the field line.

Under an earthward-directed parallel electric force an interesting situation may arise for particles injected from P toward the equator ($v_{\parallel i} < 0$), i.e., away from

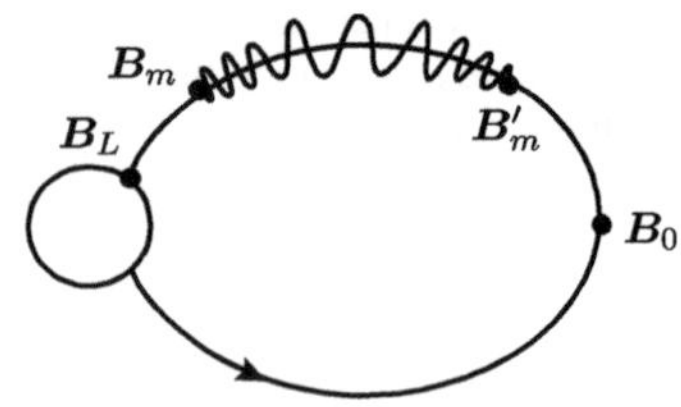

Fig. 2.8 Case of a particle trapped between two mirror points in the same hemisphere, for certain conditions of the field-aligned electric field

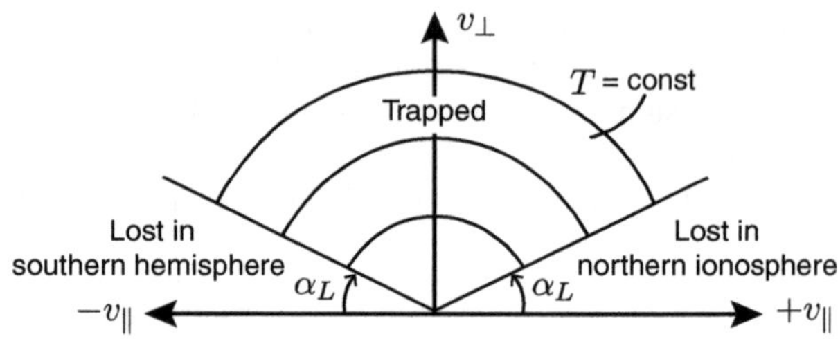

Fig. 2.9 Background of a $v_\perp, v_\parallel$ map, for charged particles at a given point P of a centered dipole field line, showing loss-cone angle and constant kinetic energy contours. $+v_\parallel$ points to the northern hemisphere

the ionosphere. An examination of relation (2.31) shows that for certain V's and T_i's, the correction function may win over the other term $(\sin^2\alpha_i/B_i)B(s)$, and the particle may reach a mirror point *before* it crosses the equatorial point of the field line. In other words, a particle may be trapped on the *same* hemispheric side of a field line (see sketch in Fig. 2.8)! This indeed can happen to electrons under auroral electric field conditions. Positive ions, on the other hand, traveling under these electric field conditions upwards from the ionosphere decrease their pitch angles; bundles of ions from the upper ionosphere accelerated and focused along the field line because of this process are called "conics". In Sect. 3.3 we shall mention a region of the dayside magnetosphere near the boundary where the magnetic field has a slight secondary *maximum* on the equator, where particles can be trapped in high-latitute pockets on the same side of the equator.

A useful representational device are the so-called velocity maps in which a guiding center particle is represented as a point, and on which directional particle fluxes (see Sect. 4.1) can be mapped and key regions and their delimitations identified. For instance, in absence of any external forces, the velocity map of a particle injected at a point P of the field line would look as shown in Fig. 2.9. This figure is drawn for a given guiding center position on a field line of a pure dipole, and shows constant kinetic energy T and pitch angle α contours; trapping and loss regions are also indicated. One point on the map represents a particle with a given pair of velocities $v_\perp, v_\parallel$ at that guiding center position on the field line.

This map becomes more interesting when the field line is no longer an equipotential. Let us combine (2.29) and (2.30) into the following equation, for the case of a field-aligned electrostatic potential $V(s)$:

$$v_\parallel^2(s) = v_{\parallel i}^2 + v_{\perp i}^2\left[1 - \frac{B(s)}{B_i}\right] - \frac{2q}{m}\left[V(s) - V_i\right] \tag{2.44}$$

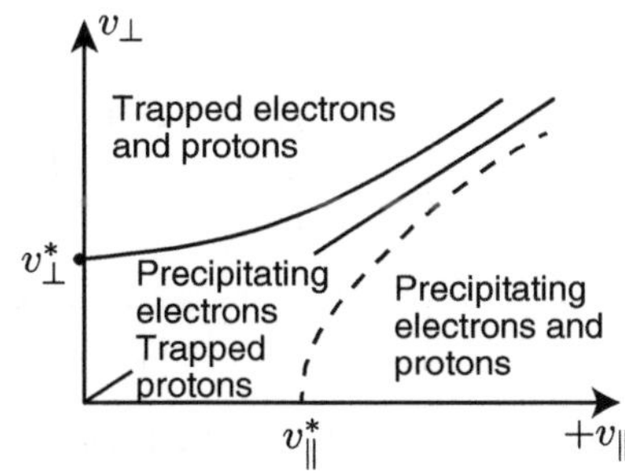

Fig. 2.10 Limiting curves for earthward moving electrons and ions on a $v_\perp$-$v_\parallel$ map

$v_\parallel^2(s)$ will now be equal to zero at mirror points due to the combined action of a mirror force and an electric field force; these points always separate an allowed segment of the field line from an off-limits one. In our example, for particles starting from s_i toward the Earth, we have $B(s) > B_i$ and $qV(s) > qV_i$; for particles starting toward the equator, reverse the inequalities.

The regions of the velocity map representing particles which *either* mirror or precipitate will be separated by a curve of points $v_\perp$, $v_\parallel$ for which at s_L the value of (2.44) is $v_\parallel^2(s_L) = 0$. For electrons, the equation of this separatrix (Fig. 2.10) is given by

$$v_\perp = \sqrt{\frac{v_\parallel^2 + 2|q|/m(V_L - V_i)}{(B_L/B_i - 1)}} \tag{2.45}$$

Notice in the figure the two regions separated by the hyperbola branch given by this equation: (i) the upper region of initial $v_\perp$, $v_\parallel$ values, for which the electron will mirror as it travels toward the Earth before it precipitates into the ionosphere at L; (ii) the lower region of initial values for which the electron accelerated by the electric field will precipitate. The asymptote of the hyperbola branch shown represents the loss cone in absence of any field-aligned electric field. Note that even electrons with an initial 90° pitch angle ($v_\parallel = 0, v_\perp = v_\perp^*$) can be drawn into the ionosphere and precipitate. For positive ions, Eq. (2.45) will have a negative sign in front of $2|q|$ and the separatrix is a conjugate hyperbola, as shown in broken line in Fig. 2.10. Ions that would have been lost in absence of an electric field will now mirror before reaching the ionosphere as the result of a concerted action of electric and mirror forces.

Our next task is to examine what will happen to electrons initially traveling in the opposite direction, i.e., toward the equator (decreasing s, $v_{\parallel i} < 0$). Those are the particles which, as mentioned above, will be decelerated by the external force and may have a chance, under specific conditions, of mirroring *before* reaching the equator, i.e., of being trapped within one hemispheric branch of the field line. The separatrix between these two regions is given by the curve that corresponds to initial $v_\perp$ and $v_\parallel$ values for which the electron mirrors exactly *at* the minimum-B point, i.e., at B_0. Replacing $B(s)$ and $V(s)$ in (2.44) by B_0 and V_0, respectively, and taking

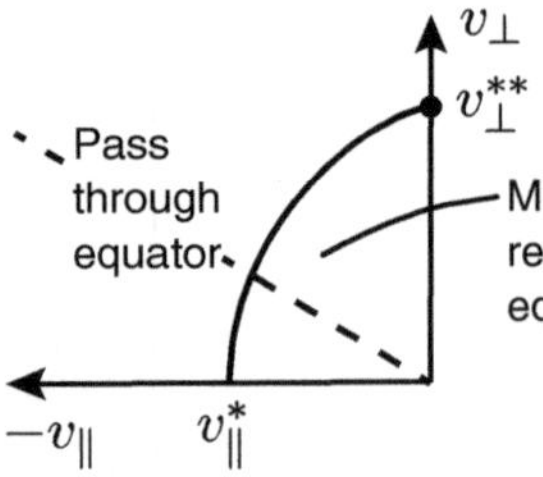

Fig. 2.11 Same as Fig. 2.10, for equatorward moving particles

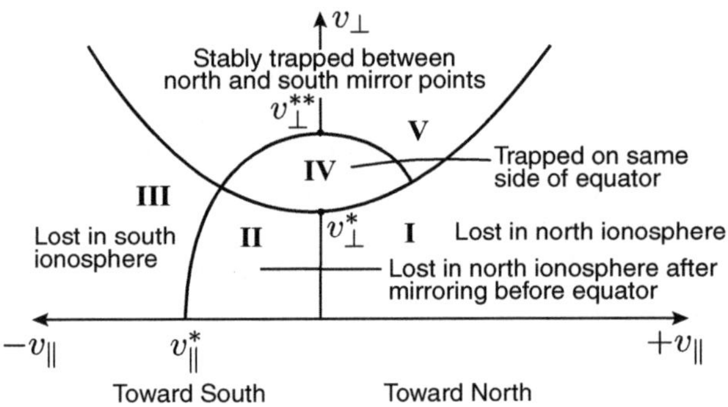

Fig. 2.12 Sketch of characteristic regions in a $v_\perp, v_\parallel$ map, for electrons injected at a northern hemisphere point of a centered dipole field line with an ionosphere-to-equator directed parallel electric field. The separatrix curves are given by (2.45) (hyperbola) and (2.46) (ellipse), respectively. For explanation, see text

into account that now $B(s) < B_0$ and $V(s) < V_0$, we obtain for the case $v_{\parallel 0} = 0$:

$$v_\perp = \sqrt{\frac{2|q|/m(V_i - V_0) - v_\parallel^2}{(1 - B_0/B_i)}} \tag{2.46}$$

The curve is now the quadrant of an ellipse (Fig. 2.11). For initially mirroring electrons ($v_{\parallel i} = 0$) and for which $v_{\perp i} < v_\perp^{**}$, there is another mirror point *before* they reach the equator. For locally field-aligned electrons ($v_{\perp i} = 0$) for which $v_{\parallel i} < v_\parallel^{**}$, there is a mirror point, too, before they reach the equator!

It is very instructive to "play" with these velocity maps and learn how different forms of electrostatic potential, electric charge of particles, degrees of field asymmetries and initial positions and kinetic energies group the particles into different classes with respect to their parallel motion along a field line (see also Fig. 1 in [1]). As a final example, we combine the upward and downward injection cases of electrons in a symmetric dipole-like field under auroral electric field conditions into just one velocity map in Fig. 2.12. This sketch is drawn for electrons passing through a field line point in the northern hemisphere. Carefully observe the properties of five distinct classes. In a real case, regions I, II and III would be empty, except for backscattered electrons; region IV is a class all by itself, consisting of electrons trapped within the northern half of the field line. Region V includes all stably trapped electrons with mirror points in both hemispheres. It is a good exercise to draw a similar map for positive ions on auroral field lines.

Reference

1. Y.T. Chiu, M. Schulz, Self-consistent particle and parallel electrostatic field distributions in the magnetospheric-ionospheric auroral region. J. Geophys. Res. **83**, 629–642 (1978)

Chapter 3
Drift Shells and the Second and Third Adiabatic Invariants

3.1 Bounce-Average Drift Velocity and Drift Shells

When a particle bounces along a field line it also drifts perpendicularly to it with an instantaneous drift velocity given by the general expression (2.14). We shall assume this drift to be "slow": during one bounce a particle drifts less than one Larmor radius away from the initial field line, a condition that is usually fulfilled under the adiabatic conditions (2.1), (2.2) or (1.48). In previous sections, we have averaged out the particle's cyclotron motion and worked with a fictitious guiding center particle, with the electromagnetic effects of the periodic cyclotron motion mimicked by the action of an invariant quantity, the magnetic moment $M = T_{\perp}/B$.[1] There are many situations, especially in radiation belt physics, in which it is not necessary to keep track of the periodic bounce motion up and down the field line, or "bounce phase" (conveniently represented by the field line's arc length s). In such a case, the second periodicity characterized by the bounce period (2.34) can also be averaged out, leaving us with the concept of a "bare" drift motion of the field lines successively occupied by the bouncing guiding center particle, thus generating a surface (Fig. 3.1). We may call the field line along which a particle is bouncing at any given time the *guiding field line*. Each guiding field line is limited by the particle's mirror points. The surface generated by the guiding field line is called a *drift shell*. The mirror points generate the two limiting curves m, m' called *mirror point traces* (or *rings* if the surface is closed). To define the geometric features of a drift shell we need to: (i) identify the initial guiding field line; (ii) find the velocity with which the particle is changing guiding field lines; and (iii) integrate this velocity, thus identifying the subsequent guiding field lines.

[1] In the case of radiation belt particles, it is safe to neglect in (1.6) the drift velocity V_D, therefore it is also safe to replace the GCS velocity $v_{\perp}{}^*$ in the definition of the magnetic moment (1.24) by the transverse velocity $v_{\perp}$ in the OFR. We shall do so until further notice.

J.G. Roederer and H. Zhang, *Dynamics of Magnetically Trapped Particles*, Astrophysics and Space Science Library 403, DOI 10.1007/978-3-642-41530-2_3,

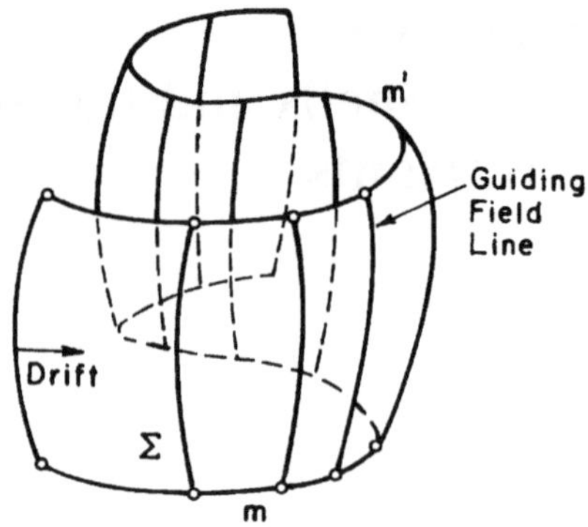

Fig. 3.1 Guiding field lines and generated drift shell Σ

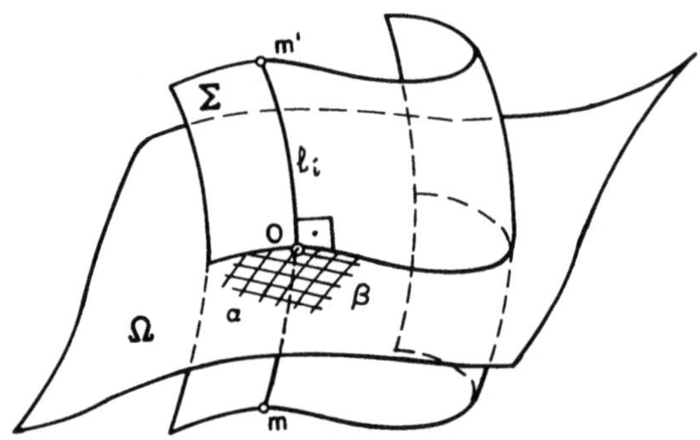

Fig. 3.2 Reference surface and coordinates of the point of a field line

There are many ways of geometrically or analytically identifying a field line in a given magnetic field. The most practical one for our purpose here is the following: Consider a given fixed reference surface $\boldsymbol{\Omega}$ (Fig. 3.2) orthogonal to all field lines; in usual trapping geometries such a surface always exists—the condition is that no field-aligned currents flow (to avoid field line torsion, see Appendix A.1), which we will assume to hold until further notice. Each field line is completely specified by the position of its intersection point O on the reference surface $\boldsymbol{\Omega}$. If α, β is a curvilinear coordinate system on $\boldsymbol{\Omega}$, the two scalars α, β specify a given field line ℓ (Fig. 3.2).[2] We already mentioned in the previous section that a particularly convenient referencein the geomagnetic field is the minimum-B equatorial surface

[2]For a divergence-free vector field like the magnetic field, the so-called *Euler coordinates* (e.g., [1, 2]) can be used, defined as $\alpha = \alpha(\boldsymbol{r}, t)$, $\beta = \beta(\boldsymbol{r}, t)$, with the property vector potential $\boldsymbol{A} = \alpha \boldsymbol{\nabla} \beta$ and therefore $\boldsymbol{B} = \boldsymbol{\nabla}\alpha \times \boldsymbol{\nabla}\beta$. A magnetic field line ℓ is defined by $\alpha = \alpha_\ell =$ const., $\beta = \beta_\ell =$ const.; the third coordinate in the Euler system can be the field line arc length s, a curvilinear distance to some reference surface $\perp \boldsymbol{B}$.

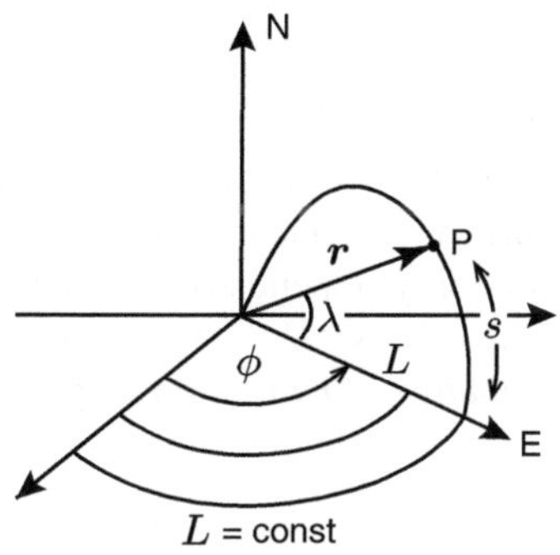

Fig. 3.3 Euler coordinates L, ϕ and s for a dipole field

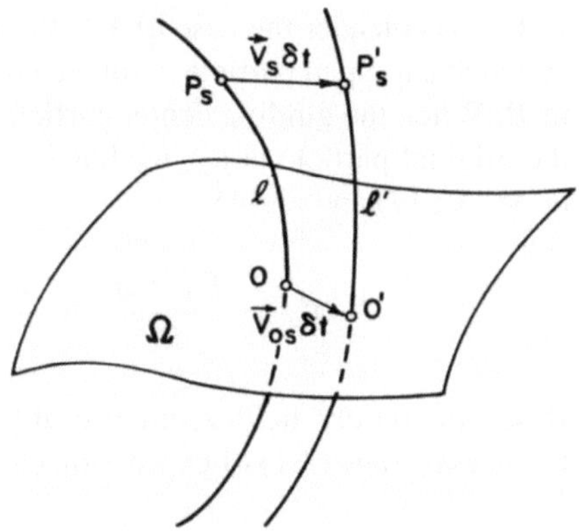

Fig. 3.4 Instantaneous and bounce-average drift velocities

(locus of the minimum-B points of all field lines and, therefore, also locus of all bounce motion equilibrium points). In the case of a pure dipole (or any field with north—south symmetry) it is a plane, in which case the two most convenient α, β parameters are proportional to the polar coordinates (radius vector r_0 and longitude ϕ_0) of the equatorial point E of the field line (see Fig. 3.3): $\alpha \sim r_0$ and $\beta \sim \phi_0$. In a dipole field, r_0 can be thought of as a 3-D scalar function $L = L(\boldsymbol{r})$, namely the modulus of the radius vector of the minimum-B point of a field line traced through *that* point $\boldsymbol{r}$. All points of a dipole field line have the same L-value (for the relations between L, s and $\boldsymbol{r}, \lambda$ see Sect. 3.4). In the general case of a more complex field (e.g., the field in a mirror machine or field lines near the dayside boundary of the magnetosphere) there may be more than one minimum-B point along one field line and the equatorial surface may have several branches (e.g., [3]). We shall not consider such a complication. As a guiding center particle bounces and drifts occupying successive guiding field lines, the intersection point O with the reference surface moves along that surface. We can introduce what is called the particle's *bounce-average drift velocity*. Refer to Fig. 3.4: if $\boldsymbol{V}_s$ is the actual, instantaneous drift velocity of the particle while its guiding center is passing through a point P at arc position s_p, the displacement of point O associated with the instantaneous displacement $\boldsymbol{V}_s\delta t$ (where $\delta t < \tau_c$) will be given by $\overrightarrow{\boldsymbol{OO'}}$ ($|OO'| \ll \rho_L$). The point O′ is obtained by tracing the field line ℓ' that goes through P′ down to the reference surface Ω. The associated drift velocity $\boldsymbol{V}_{0s} = \overrightarrow{\boldsymbol{OO'}}/\delta t$ on the reference surface is related to the actual local drift velocity $\boldsymbol{V}_s$ through a transformation that only depends on the actual field geometry (in the case of a dipole field, it is $V_{0s} = \cos^{-3}\lambda V_s$, where λ is the latitude of point P). The *bounce-average drift velocity* of O will then be given by:

$$\langle \boldsymbol{V}_0 \rangle = \frac{1}{\tau_b}\int_0^{\tau_b} \boldsymbol{V}_{0s} dt = \frac{2}{\tau_b}\int_{s'_m}^{s_m} \boldsymbol{V}_{0s}\frac{ds}{v_\parallel} \tag{3.1}$$

Let us consider the case of a static magnetic field without field-aligned currents, in which a trapped particle is subjected to an electrostatic field always perpendicular to $\boldsymbol{B}$. When the guiding center particle is at position s and the pitch angle cosine of the original particle is μ_s, we have for the local drift velocity, taking into account (1.34), (2.17) and (1.14):

$$\boldsymbol{V}_s = \frac{\boldsymbol{E}_s \times \boldsymbol{B}_s}{B_s^{\;2}} + \frac{m v_s^{\;2}}{2q B_s^{\;3}}(1+\mu_s^{\;2})\boldsymbol{B}_s \times \nabla_\perp B_s \tag{3.2}$$

Based on purely field-geometric arguments we prove in Appendix A.2 that the bounce-average (3.1) of expression (3.2) is given by

$$\langle \boldsymbol{V}_0 \rangle = \frac{\nabla_0 J \times \boldsymbol{e}_0}{q\tau_b B_0} \tag{3.3}$$

in which

$$J = \oint p_\parallel ds = m \oint v\,\mu\, ds \tag{3.4}$$

is an integral taken along the guiding field line for a complete bounce cycle ($\oint = 2\int_{s'_m}^{s_m}$). In the specific case of *equipotential field lines*, v is constant during one bounce, and taking into account the definition of the purely field-geometric quantity I (2.37),

$$J = 2pI \tag{3.5}$$

In Appendix A.2 we also show that for equipotential field lines (p is constant along a field line, but not across it!):

$$\nabla_0 J = 2p\nabla_0 I + q\,\tau_b\,\boldsymbol{E}_0 \tag{3.6}$$

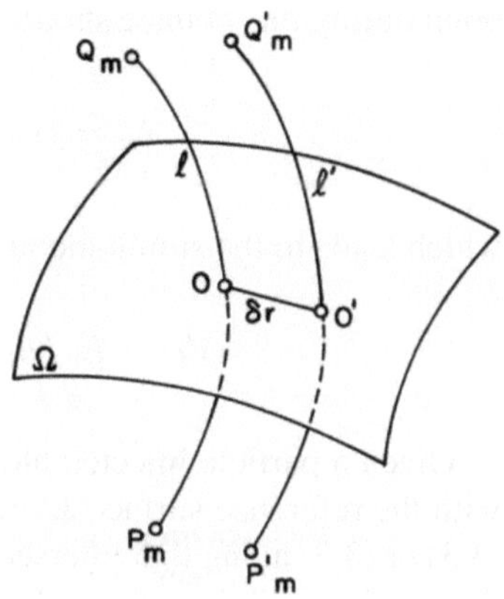

Fig. 3.5 Field line configuration for the definition of the gradient of J or I

With this relation, (3.3) becomes, replacing τ_b with $2S_b/v$:

$$\langle \boldsymbol{V}_0 \rangle = \frac{\boldsymbol{E}_0 \times \boldsymbol{B}_0}{{B_0}^2} + \frac{mv^2}{qS_b B_0} \nabla_0 I \times \boldsymbol{e}_0 \tag{3.7}$$

All quantities in (3.3) and (3.7) are defined at the reference point O. These expressions thus represent the average drift velocity *of the reference point* of the guiding field line. In principle, since the surface Ω could be any surface normal to the field lines, that reference point could be *any* point of a field line between the particle's mirror points. Choosing for instance one of the latter, (3.7) would then become the average drift velocity of one of the particle's mirror points.

Although everybody knows how a gradient is defined, it is prudent to give the "recipe" for the numerical calculation of the gradients of J or I in the above expressions (see Fig. 3.5). It consists of the following steps: (i) Consider a particle of given magnetic moment (or kinetic energy) and pitch angle, for which $\nabla_0 J$ (or $\nabla_0 I$) is wanted (Ω is a reference surface perpendicular to all field lines). (ii) Compute numerically the integral (3.4) (or (2.37)) for the guiding field line between the mirror points (given through M and α). (iii) Take a neighboring field line ℓ' and compute J or I between the mirror points that would result for a particle with the *same* magnetic moment and total energy placed on that field line (same B_m if there are no forces). (iv) Form the ratio $\delta J/\delta r = (J' - J)/|OO'|$. (v) Repeat the process for other neighboring field lines (chosen in some "logical" way) until the one for which $|\delta J/\delta r|$ is maximum is found. The direction of $\nabla_0 J$ will be that of the vector $\overrightarrow{OO'}$; its magnitude will be the limit of $\delta J/\delta r$ for $\delta r \to 0$. Notice carefully that just like $\langle \boldsymbol{V}_0 \rangle$, $\nabla_0 J$ or $\nabla_0 I$ depend on the reference surface (point O).

With (3.7) we can find specific adiabatic conditions required for the validity of all preceding definitions. We stated before that the drift displacement of the reference point during one bounce should be much less than one typical Larmor radius (1.22):

$$\delta_0 = \langle V_0 \rangle \tau_b = \frac{E_0}{B_0}\tau_b + \frac{2p\nabla_0 I}{qB_0} \ll \rho_C$$

which leads to the simultaneously required conditions

$$V_E = E_0/B_0 \ll \frac{\rho_C}{\tau_b} \sim v\frac{\tau_C}{\tau_b} \quad \text{and} \quad \nabla_0 I \ll 1 \tag{3.8}$$

Given a particle injected along an initial field line, one can find the intersection with the reference surface Ω of all subsequent guiding field lines by integration of (3.3) or (3.7) along that intersection curve. The full drift shell is then determined by tracing the field lines through all reference points up and down to the local mirror points defined by the conservation of M and $\mathfrak{E}$. This may be quite lengthy; in many practical applications one just wants to find the guiding field line at some given longitude far away from the initial field line, without having to trace the intermediate drift track. To accomplish that, we can use the fundamental property of J (or I for a static magnetic field without external forces) as a second adiabatic invariant and search for the particle's mirror point at the wanted longitude—as we shall show in the next section.

3.2 The Second Adiabatic Invariant

The quantity J (3.4) is an adiabatic invariant conserved during the drift of a guiding center particle in a trapping geometry,[3] provided adiabatic conditions (2.1), (2.2) and (3.8) apply. This conservation theorem follows at once from (3.3): since $\langle \boldsymbol{V}_0 \rangle$ is of the form "scalar times $(\nabla_0 J \times \boldsymbol{e})$", the drift is in a direction *normal* to the gradient of J, so that on the following guiding field lines the value of J will remain constant. Remember that J must be computed for constant M and $\mathfrak{E}$, and that the mirror points must be determined through (2.38) or (2.43).

Let us first consider the case in which *no external forces are acting and the magnetic field is static*. Conservation of J implies that of the field-geometric integral I (2.37). Furthermore, since the potential energy W and the particle's kinetic energy T are constant in this case, the conservation of magnetic moment $M = T_\perp/B = T/B_m$ leads to the conservation of mirror field intensity. In summary, we have

[3]Going back to the footnote on page 10 concerning canonical path integrals, we can consider the bounce motion as a cyclic motion, with the field line arc length s as the variable. In that case, $J_b = \oint (\boldsymbol{p} + q\boldsymbol{A}) \cdot \boldsymbol{ds}$. This time the cyclic orbit (up and down the field line) encompasses *zero* magnetic flux; therefore only the first term subsists and is equal to $J_b = \oint \boldsymbol{p} \cdot \boldsymbol{ds}$, equal precisely to what we defined as the second invariant J, which is therefore adiabatically constant.

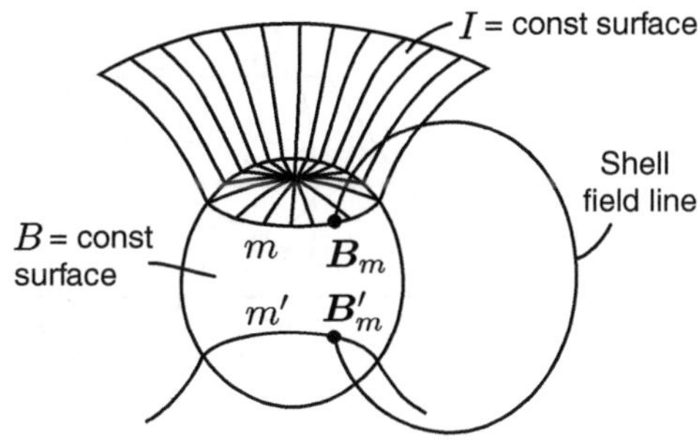

Fig. 3.6 Sketch of the geometric relations between B =const., I =const. and drift shell surfaces

$$I = \text{const.} \tag{3.9}$$

$$B_m = \text{const.} \tag{3.10}$$

These relations define the drift shell of a particle as the surface generated by all field lines that pass through the intersection curves m, m' (mirror traces) of the corresponding I = const. and B_m = const. surfaces (Fig. 3.6). The values of I and B_m are determined by (2.37) and (2.33), respectively, which in turn are calculated for the initial position and pitch angle of the particle in question. In all this, the second invariant has to be envisioned as a *scalar function of space* $I = I(\boldsymbol{r})$ defined by the I-value of a particle mirroring *at that point* $\boldsymbol{r}$. Relations (3.9) and (3.10) tell us that for a static case in absence of external forces, drift shells are independent of a particle's mass, charge and energy: they only depend on the particle's initial guiding center position and pitch angle. Note carefully that the constancy of I and B_m for a drifting particle does not imply constancy of the bounce path (2.35) or bounce-average drift velocity, which in this special case is

$$\langle \boldsymbol{V}_0 \rangle = \frac{mv^2}{qS_b B_0} \nabla_0 I \times \boldsymbol{e}_0 \tag{3.11}$$

In a pure dipole magnetic field we do not have to invoke the conservation of the second invariant I: for reasons of azimuthal and N-S symmetry, it is enough to determine the field line (given by the parameter L, see Fig. 3.3) through the initial position and the mirror point B_m given by (2.33), and then rotate the portion of field line between conjugate mirror points around the dipole axis.

Unfortunately, for any other field geometry with no symmetries, no simple analytical expressions exist that can be used to determine the drift shell field lines of a trapped particle. The integral function $I = I(\boldsymbol{r})$ must be determined *numerically*, using a mathematical model (numerical or analytical) of the magnetic field $\boldsymbol{B} = \boldsymbol{B}(\boldsymbol{r})$. Given a particle injected with a certain pitch angle at a point of some initial field line, the computational procedure to find the particle's shell field line at any other longitude is then as follows. First, once you have the initial point B value, determine B_m and compute I by tracing the field line between the two initial conjugate mirror points. Then go to the wanted longitude and by picking different points for which $B = B_m$ (see Fig. 3.6) instruct the computer to find by an iterative method *that* field line for which the value of (2.37) between mirror

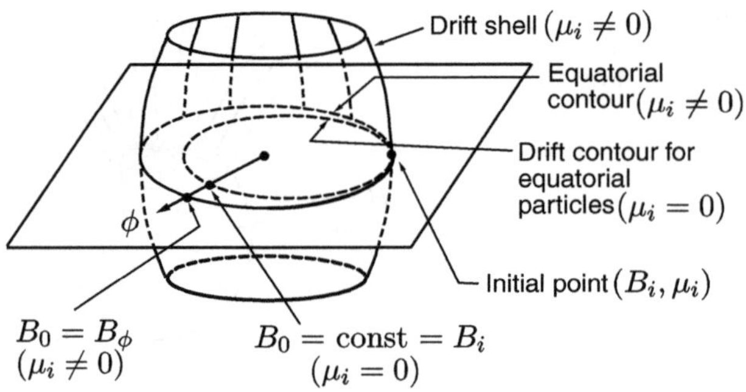

Fig. 3.7 Schematic view of the drift contour of an equatorial particle ($\mu_{0i} = \cos\alpha_{oi} = 0$) and of the drift shell of a $\mu_{0i} \neq 0$ particle, in an azimuthally asymmetric field

points is equal to the initial value $I \pm$ prefixed error. If no such field line can be found, it means that on that meridian no trapping is possible for a particle of the prescribed I, B_m values—the particle must have left its trapping region before drifting to the longitude in question. Notice that in this way we indeed can find shell field lines at arbitrary longitudes *without* having to proceed in small longitudinal drift steps from the initial guiding field line. Of course, if the drift time from initial to final longitude is wanted, it is necessary to integrate (3.11): $\tau_D = \oint \langle V_0 \rangle^{-1} d\ell$.

For particles mirroring close to the equator, we can use an expansion of $B(s)$ similar to the one leading to (2.39) and (2.40) in the definition of I (2.37), to obtain to first order:

$$I \cong \frac{\pi}{\sqrt{2}} \left(\frac{B_m}{a_0}\right)^{1/2} \left(1 - \frac{B_0}{B_m}\right) \cong \frac{\pi}{\sqrt{2}} \left(\frac{B_0}{a_0}\right)^{1/2} \mu_0^2 \tag{3.12}$$

where $a_0 = \partial^2 B/\partial s^2$, B_0 and μ_0 are taken at the minimum-B point. This expression and the constancy of B_m and I lead us to an analytical expression to first order in ${\mu_0}^2$ of the equatorial trace (in terms of equatorial B-values) of a drift shell for near-90° pitch angle particles (Fig. 3.7):

$$B_0(\phi) = B_{0i} \left[1 - \left(\frac{a_0(\phi)}{a_{0i}}\right)^{1/2} \mu_{0i}^2\right] \tag{3.13}$$

For $\mu_{0i} \neq 0$, the second derivative $a_0(\phi) = \partial^2 B/\partial s^2|_{0\phi}$ only needs to be evaluated at the $B_0(\phi) = B_{0i} = $ const. contour (see Fig. 3.7), since under the present approximation μ_{0i}^2 is very small. Clearly, for equatorial 90° pitch angle particles as well as for azimuthally symmetric fields, the equation of the shell intersection is $B_0(\phi) = B_{0i} = $ const., as we have seen for a special case in Sect. 1.6. The longitudinal variation of $B_0(\phi)$, i.e., the azimuthal distortion of the drift shell, is determined by the longitudinal variation of $a_0(\phi)$: in all azimuthally asymmetric fields, particles with different initial pitch angles will generate different drift shells. This is called shell-splitting and will be discussed in more general terms and without pitch angle restrictions in the next section.

We end this section with a return to the more realistic situation of particle trapping in a static magnetic field *under the action of external forces perpendicular to the magnetic field lines* (i.e., still excluding field-aligned forces). Assuming that the external force derives from an electrostatic potential $\boldsymbol{F} = -q\nabla V(\boldsymbol{r})$, field lines will be equipotentials, each with a potential which we denote by $V_0(\boldsymbol{r}_0)$, the potential of its intersection with the minimum-B surface. Under such conditions, the integral I (2.37) will no longer be an adiabatic invariant; however, we can easily derive one, which we call K, by combining the invariance of the magnetic moment M (1.26) with (3.5):

$$K = \frac{J}{2\sqrt{2mM}} = I\sqrt{B_m} = \int_{s'_m}^{s_m} (B_m - B(s))^{1/2}\, ds = \text{const.} \tag{3.14}$$

This quantity still is "purely field-geometric" like I, independent of the particle—but only to a certain point: the limits of the integral in (3.14) are not constant (invariants) like those in (3.9), but depend on the initial mirror points and the local kinetic energy of the particle, which is no longer constant. The conservation of total energy $\mathfrak{E}$ and the adiabatic constancy of M and K yield the following equations that allow us to determine a particle's guiding field line at any longitude ϕ, thus if we wish, the entire drift shell (subindex i stands for initial):

$$\mathfrak{E} = T(\phi) + qV(\phi) = T_i + qV_i = \text{const.} \tag{3.15}$$

$$M = \frac{\mathfrak{E} - qV(\phi)}{B_m(\phi)} = \frac{T_i}{B_{mi}} = \text{const.} \tag{3.16}$$

$$K = I_\phi\sqrt{B_m(\phi)} = I_i\sqrt{B_{mi}} = \text{const.} \tag{3.17}$$

The numerical determination of drift shells for given models of $\boldsymbol{B}(\boldsymbol{r})$ and $V_0(\boldsymbol{r}_0)$ is qualitatively similar to the procedures given above for a magnetic field in absence of an electrostatic field, except that now the mirror point B-values are energy dependent and have to be recalculated for each field line using (3.16), and that at the wanted longitude the iterative process must zero-in on a prefixed K value ($\pm$ prefixed error) instead of I.

3.3 Shell Splitting and Pseudo-trapping

It is important to analyze in general terms some characteristic geometric properties of trapped particle drift shells in typical asymmetric magnetospheric field models (in absence of electric fields). Take a particle that starts at a given longitude ϕ_0, circling and bouncing about a guiding field line between conjugate mirror points at a field value B_m. The integral (2.37) has a value I. This means that when drifting through any other longitude, for example 180° away, this particle will bounce along a field line that passes through the intersection of the corresponding $I = $ const. and $B_m = $ const. surfaces (Fig. 3.8). Now take a particle which starts on the *same* initial

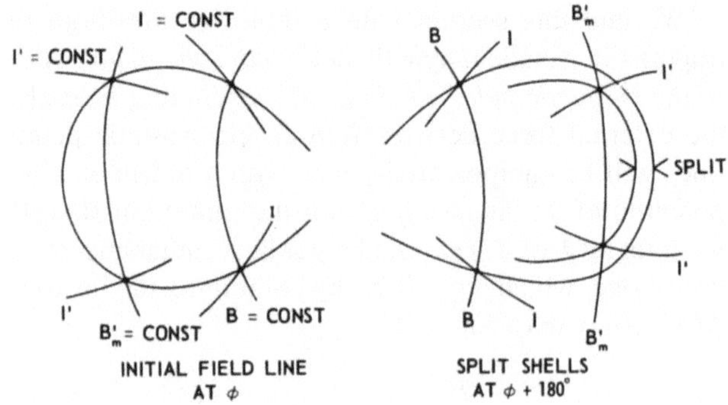

Fig. 3.8 Geometric depiction of shell splitting in an asymmetric field

field line but mirrors at a lower value, $B_m' < B_m$. Its integral I' will also be smaller, $I' < I$. After a 180° drift this second particle will be bouncing along a field line that passes through the intersection of the surfaces $I' = \text{const.}$ and $B_m' = \text{const.}$ *Only* in the case of perfect azimuthal symmetry (as in a pure dipole) will these surfaces intersect *exactly* on the same field line as that of the first particle, and thus the shells of both particles will be coincident. This latter case is called *shell degeneracy*. In the general case, however, as we already have seen for the near-equatorial case in (3.13), particles starting on a common field line at a given longitude will populate *different* shells according to their initial mirror point fields or, which is equivalent, according to their initial equatorial pitch angles. This effect is called *shell splitting* [4]. Of course, all these drift shells will be tangent to each other on the initial field line.

Let us analyze a quantitative example. In doing so, we will not insist in the accuracy of the magnetospheric field model used; the principal conclusions depend on broad features, common to all specific models. We will use some of the old, analytically simple and computationally fast models of the 1960s and 1970s, such as the Mead-Williams model [5]. Figure 3.9 shows how particles, starting from a common field line in the *noon* meridian with equatorial pitch angle cosines μ_0 of 0.2, 0.4, 0.6, 0.8 and nearly 1 (mirroring close to the Earth's surface), do indeed drift on different shells which intersect the midnight meridian along different field lines. The dots represent the particle's mirror points. Curves giving the position of mirror points for constant equatorial pitch angles are shown for comparison (in a dipole field, they should be constant latitude lines, Appendix A.2). Notice the change (decrease) in equatorial pitch angle for the same particle when it drifts from noon to midnight. Figure 3.10 depicts the same features for particles starting on a common field line in the *midnight* meridian. In this case, again, the equatorial pitch angle changes considerably when the particle drifts to the opposite meridian (increasing at noon). Notice from these two examples that, as equatorial pitch angles increase, shell splitting is directed radially earthwards for particles starting on the same field line at noon, and radially outwards for particles starting on a common field line at midnight. Furthermore, shell splitting is maximum for particles mirroring close to the equator. The more distorted the magnetospheric field (compared to a dipole-like field), the more pronounced these effects are.

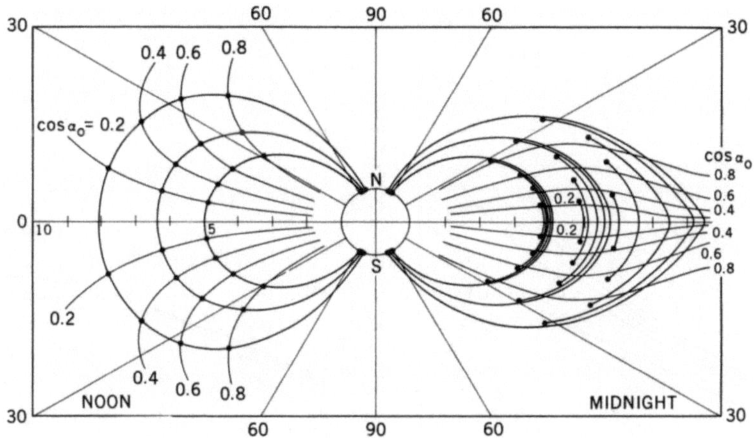

Fig. 3.9 Computed shell splitting for particles starting on common field lines in the noon meridian (a simple magnetospheric field model was used [6]). *Dots* represent particles' mirror points. *Curves* giving the position of mirror points for constant equatorial pitch angle α_0 are shown

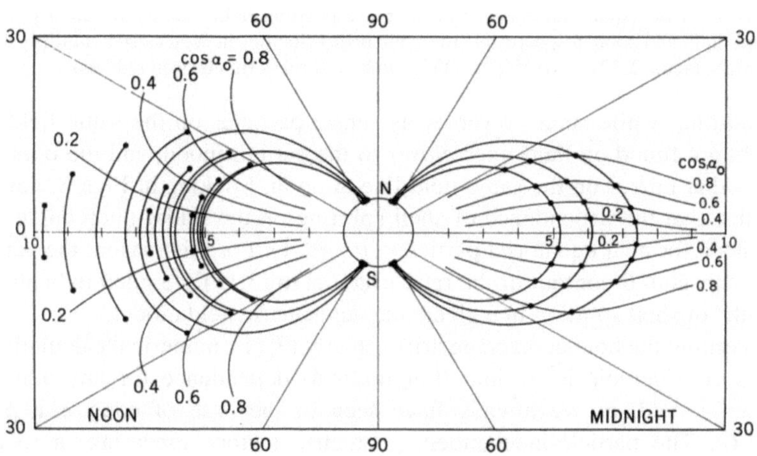

Fig. 3.10 Same as Fig. 3.9, for particles starting on common field lines in the midnight meridian

An interesting feature (not shown in Figs. 3.9 and 3.10) arises with particles of equatorial pitch angles smaller than ~45°–50° ($\mu_0 \sim 0.45$), i.e., mirroring closer and closer to the earth. For that "magic" pitch angle[4] drift shells are azimuthally symmetric (like in a dipole); as the pitch angle decreases further (μ_0 increases),

[4]Magic, because it is practically independent of magnetospheric field model details and of the radial distance to the field line's equatorial point!

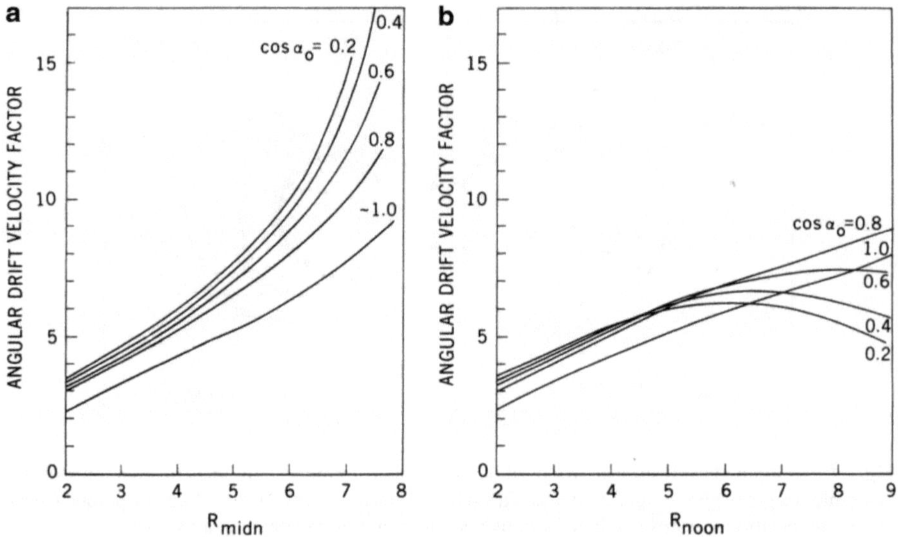

Fig. 3.11 Value of the geometric factors intervening in the bounce-average angular drift velocity as a function of the equatorial distance of the corresponding field line, for different pitch angles and or midnight and noon, respectively. To obtain drift velocities in degrees/sec, multiply the above values by the factor $2.321 \times 10^{-2}(\gamma^2 - 1)/\gamma$ times rest mass (in electron masses)

shell splitting, while small, reverses its sense: particles on the same field line at midnight are found on field lines *closer* to the earth at noon, and the opposite for particles that mirror on the same field line at noon. Looking at Figs. 3.9 and 3.10, we realize that the dependence of shell splitting on the initial pitch angle is most pronounced for near equatorial particles ($\mu_0^2 \ll 1$). For this reason, the variability along a constant-B contour of the ratio $a(\phi)/a_0$ in (3.13) is a good indicator of the "strength" of shell splitting in a given magnetospheric field model.

Concerning the bounce-average drift velocity (3.11), numerical calculations also reveal a considerable local-time (longitudinal) dependence for any azimuthally asymmetric field, as we already have seen in the case of equatorial particles (Sect. 1.6). The particle-independent geometric factors appearing in (3.11) are represented in Fig. 3.11 as a function of the equatorial distance of the corresponding field line, for different pitch angles and for the midnight and noon meridians, respectively. For better understanding *angular* drift factors are shown. For radial distances $\lesssim 3\,R_E$ we observe a dipolar dependence. Beyond $\sim 3\,R_E$ there is a considerable departure. Angular drift velocities on the night side are appreciably higher than on the dayside. The peculiar inversion of the pitch angle dependence occurring on the dayside is due to the fact that particles mirroring at low latitudes experience on the average during their bounce a greatly reduced field gradient, whereas those mirroring at high latitudes spend a larger fraction of their bounce-time in a dipole-like field and hence drift faster. A closer inspection of the local-time dependence of the bounce-average drift velocity (not shown) shows that a given particle trapped in the outer magnetosphere ($r \gtrsim 6\,R_E$) spends from 2/3 to 3/4 of its time on the dayside. In other words, there is always a higher probability of finding a trapped particle on the dayside than on the night side. This has important consequences for trapped particle diffusion.

When a field line has its equatorial point beyond about 8 R_E, a fraction of the particles mirroring on it are *pseudo-trapped* or *quasi-trapped*, being unable to complete a drift around the Earth. In particular, as we already have seen for equatorial particles in Sect. 1.6, particles mirroring at midnight close to the minimum-B surface, will abandon the magnetosphere through the boundary near dawn or dusk (depending on their charge, i.e., the direction of drift) before reaching the noon meridian. On the other hand, particles mirroring at high latitudes on the dayside will run into the magnetospheric tail before reaching the midnight meridian. Figure 3.12 shows computed limits between stable trapping and quasi-trapping regions on the noon-midnight meridian. At other local times one quasi-trapping region disappears at the expense of the growth of the other. The calculation of open drift shells in the pseudo-trapping regions using the adiabatic invariants is cumbersome; recently, full particle orbits in the high-latitude pseudo-trapping region were computed by integrating numerically Eq. (1.1) (Fig. 3 in [7]).

The theoretical boundary of stable trapping is thus a surface that separates two regions in which lie the mirror points of particles whose drift shells are either closed around the Earth or open, respectively. Whatever particles one finds mirroring and drifting inside the regions of quasi-trapping must have been injected from elsewhere; this is why one should expect low and fluctuating fluxes in the regions of quasi-trapping, as opposed to a larger, more smoothly varying flux in the region of stable trapping. The regions of quasi-trapping are limited on their outer side by the boundary of closed field lines, along which no adiabatic trapping is possible. An equivalent way of characterizing the limit of stable trapping is with the concept of a *drift loss cone*. At the minimum-B equator, this cone encompasses all pitch angles for which a particle mirrors inside the region of quasi-trapping. On the night side, the drift loss cone is oriented perpendicularly to the local field line; on the dayside, it is directed along the field line. Figure 3.13 shows the drift loss cones in addition with the "conventional" *bounce loss cones* along the field line.

Much of the preceding results obtained through numerical computer calculations using a model magnetospheric field can be described in a more approximate, but still physically meaningful, analytical way if one restricts the discussion to near-equatorial particles ($\cos\alpha \ll 1$) and uses a simple first-order analytical approximation of the magnetospheric field, as we did in Sect. 1.6. Neglecting any electric currents in the region to be examined (say, between 3 and 5 R_E during quiet times), we shall consider a field [8] derived from the following magnetic scalar potential (which gives the equatorial magnetic field (1.49) used in Sect. 1.6):

$$V(r,\theta,\phi) = R_E\left[-B_E\left(\frac{R_E}{r}\right)^2 + b_1\left(\frac{r}{R_E}\right)\cos\theta + \frac{1}{2}b_2\left(\frac{r}{R_E}\right)^2\sin(2\theta)\cos\phi\right] \tag{3.18}$$

The coefficients b_1 and b_2 are those of relations (1.50).

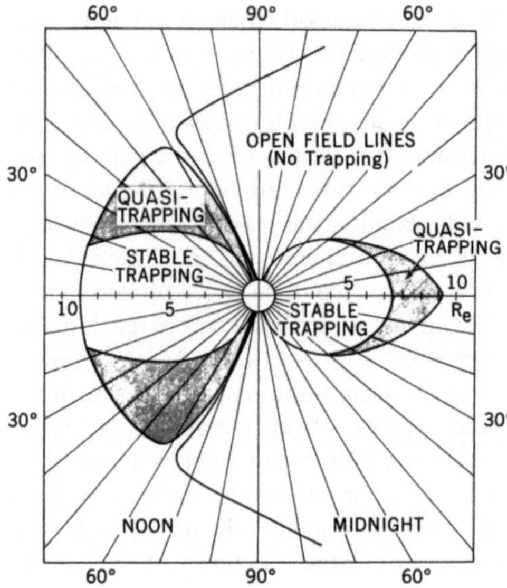

Fig. 3.12 Location of the quasi-trapping regions in the magnetosphere. Particles mirroring inside those regions are unable to complete a 360° drift around the earth. Those injected into the *left side* will be lost into the tail; those injected into the *right portion* will abandon the magnetosphere through the boundary on the dayside

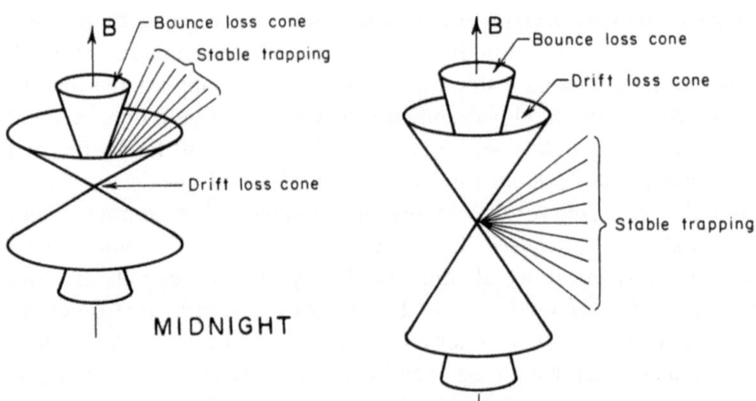

Fig. 3.13 Drift and bounce loss cones on the midnight and noon meridians

For convenience, we reproduce here the expression (1.49) for the magnetic field intensity B_0 on the equatorial plane:

$$B_0 = B_E \left(\frac{R_E}{r_0}\right)^3 \left[1 - \frac{b_1}{B_E}\left(\frac{r_0}{R_E}\right)^3 - \frac{b_2}{B_E}\left(\frac{r_0}{R_E}\right)^4 \cos\phi_0\right]$$

The quantity $a_0 = \partial^2 B/\partial s^2$ at a given equatorial point r_0, ϕ_0 can be (laboriously) calculated to obtain [9]:

$$a_0(r_0, \phi_0) = \frac{9B_E}{{r_0}^5}\left[1 + 2\frac{b_1}{B_E}\left(\frac{r_0}{R_E}\right)^3 + \frac{29}{8}\frac{b_2}{B_E}\left(\frac{r_0}{R_E}\right)^4 \cos\phi_0\right] \tag{3.19}$$

Combination of the two latter equations with (3.12) and (3.13) leads to the equation of the equatorial trace of the drift shell generated by a particle injected at point r_{0i}, ϕ_{0i} with a pitch angle cosine μ_{0i} ($\ll 1$):

$$r_0 = r_{0i} - \frac{R_E}{3}\frac{b_2}{B_E}\left(\frac{r_{0i}}{R_E}\right)^5 \left(1 - \frac{43}{18}{\mu_{0i}}^2\right)(\cos\phi_0 - \cos\phi_{0i}) \tag{3.20}$$

Compare this with (1.51); again, only the day-night asymmetry coefficient b_2 appears. An analysis of the case of equatorial particles ($\mu_{0i} = 0$) was given in Sect. 1.6.

According to (3.19) and for field parameters with $R_s = 10$, the quantity $a_0 = \partial^2 B/\partial s^2$ goes through zero at ${r_0}^* = 6.9\,R_E$ on the noon meridian ($\phi_0 = \pi$), becoming negative beyond. As one moves away from the noon meridian toward dawn or dusk, the critical distance ${r_0}^*$ increases and approaches the magnetospheric boundary. This obviously means that beyond a certain distance on the dayside, the $\theta = \pi/2$ plane *is no longer a minimum-B surface*. We already mentioned (page 59) that in that region, field lines attain their minimum B-value at two points situated at a certain finite latitude up and down from the equatorial plane. Again, since the field approximation used here starts breaking down at these distances, more realistic field models must be used to explore numerically the actual geometry of these *minimum-B pockets* and their effects on quasi-trapped particles.

Let us now examine Eq. (3.20) for $\mu_{0i} \neq 0$ (but still $\ll 1$). Drift-shell splitting becomes apparent. For a particle injected at midnight ($\phi_{0i} = 0$), the maximum radial deviation from a $\mu_{0i} = 0$ orbit occurs at noon ($\phi_0 = \pi$) and is given by

$$\Delta r_0 = r_0\big|_{\mu_{0i}=0} - r_0\big|_{\mu_{0i}\neq 0} = \frac{43}{27} R_E \frac{b_2}{B_E}\left(\frac{r_{0i}}{R_E}\right)^5 {\mu_{0i}}^2 \tag{3.21}$$

This is a positive quantity, which means that for particles with decreasing pitch angle at the midnight point (μ_{0i} increasing), drift shells will reach out farther on the dayside (see Fig. 3.10). For a particle injected at noon, the situation at midnight is reversed (the above equation has a minus sign); for smaller pitch angles at noon, the shells are displaced inward (toward the earth) at midnight. Notice the strong dependence on radial distance r_{0i}.

An interesting fact is the appearance of the combination $[1 - (43/18)\mu_{0i}{}^2]$ in (3.20). If we forget for a moment the condition $\mu_{0i} \ll 1$, it would mean that there is a "magic" pitch angle $\alpha^* = \arccos\sqrt{18/43} = 49.6°$, a sort of "universal constant" for a field of the type given by (3.18) (i.e., independent of the actual values of the parameters B_E, b_1, b_2 and of the coordinates r_0, ϕ_0), for which drift shells are azimuthally symmetric. As we shall see in Sect. 3.5, this characteristic is still retained if one uses more refined external field models and numerical shell tracing. One consequence is that whenever one measures particles with equatorial pitch angles of about 45°–50° (mirroring at about 23° geomagnetic latitude), it really is not necessary to specify the local time (or longitude) of the measurement, in spite of the asymmetry of the magnetic field. There is a simple physical explanation for the reversal of the shell asymmetry for a certain equatorial pitch angle. When $\mu_{0i} = 0$, particles follow *constant-B* contours, which approach the earth closest on the nightside. When $\mu_{0i} \to 1$, on the other hand, conservation of the invariant I can be shown to lead to the approximate constancy of the field-line *arc length* between mirror points. Since field lines are stretched out farther on the nightside, the corresponding shells also must reach out farther on the dayside because of the more compressed field lines there. Hence there must be an intermediate value of the pitch angle for which a shell is axisymmetric. The remarkable thing is that for magnetospheric-type field geometries this pitch angle depends very little on position and field parameters.

3.4 Effects of Internal Field Multipoles on Inner Magnetosphere Particle Shells; McIlwain's L-Value

In a pure dipole field, for reasons of longitudinal and latitudinal symmetry, drift shells are degenerate and there is no shell splitting—particles on one given field line, regardless of their pitch angle, will share the same field lines as they drift; only their drift speeds will differ, depending on pitch angle. Figure 3.14 reminds the reader of the symbols used. The fundamental sets of equivalent coordinates are $(\boldsymbol{r}, \lambda)$, (L, λ) or (B, L), with the parameter $L = r_0/R_E$ (r_0: equatorial point of a field line, R_E= Earth radius $\simeq$ 6,371 km). B_L is the B-value at the field line intersection with the Earth (or ionosphere) and B_E the value of B at the equatorial point on the Earth's surface (proportional to the Earth's dipole moment: $B_E = k_0\, R_E^{-3} \simeq 0.31$ Gauss). Of course, in three dimensions we must also include the longitude ϕ. For other planets or magnetized moons, just insert the corresponding values for R_E and B_E in what follows (e.g., for Jupiter, $B_J \cong 4.28$ Gauss).

The most important relationships between those variables for what follows (see Fig. 3.14) are:

$$L = \frac{r}{R_E}\frac{1}{\cos^2\lambda} \tag{3.22}$$

$$\cos^2\lambda_L = \frac{1}{L} \tag{3.23}$$

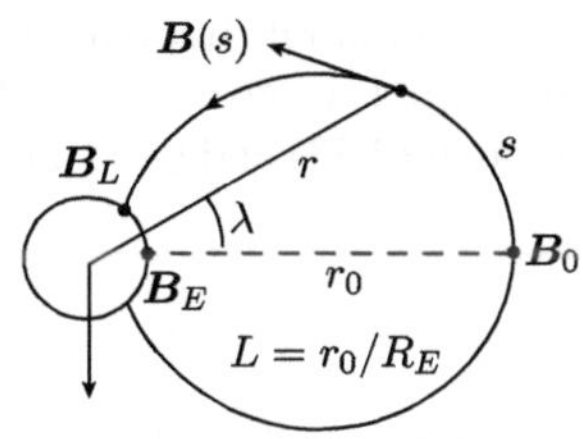

Fig. 3.14 Dipole field line parameters

$$\begin{aligned} B(L,\lambda) &= \frac{B_E}{L^3}\frac{\sqrt{4-3\cos^2\lambda}}{\cos^6\lambda} = B_0\frac{\sqrt{4-3\cos^2\lambda}}{\cos^6\lambda} && (3.24)\\ B_L &= B_E\sqrt{4-3/L} && (3.25)\\ A &= B_E R_E/(L^2\cos\lambda) \quad \text{(in the direction of } +\phi) && (3.26)\\ ds &= LR_E\cos\lambda\sqrt{4-3\cos^2\lambda}\,d\lambda && (3.27) \end{aligned}$$

For the loss cone pitch angle at any B, L point we have

$$\mu_L^2 = 1 - \frac{B}{B_E}\frac{1}{\sqrt{4-3/L}} \tag{3.28}$$

Concerning the invariant I (2.37), there is a direct relationship with the L-value and B_m for a dipole field, based on the above relations:

$$I = LR_E\int_{-\lambda_m}^{+\lambda_m}\sqrt{1-\frac{B_E}{B_m L^3}\frac{\sqrt{4-3\cos^2\lambda}}{\cos^6\lambda}}\sqrt{4-3\cos^2\lambda}\cos\lambda\,d\lambda \tag{3.29}$$

The integral limits $\pm\lambda_m$ are the latitudes of the conjugate mirror points, obtainable from the inverse function of (3.24), $\lambda_m = \lambda_m(B_m L^3/B_E)$. Expression (3.29) is of the form

$$I = LR_E h\left(\frac{B_m L^3}{B_E}\right)$$

where h is a function for which there is no closed analytical expression; it must be determined by numerical integration. Cubing and multiplying both sides by B_m/B_E, we obtain for the inverse relationship

$$\frac{L^3 B_m}{B_E} = F\left(\frac{I^3 B_m}{R_E^3 B_E}\right) \tag{3.30}$$

This defines the L-parameter as a function of the pair I, B_m. The dimensionless function $y = L^3 B_m/B_E = F(x)$ of the dimensionless argument $x = I^3 B_m/(R_E^3 B_E)$ can be found in published tables (e.g., Appendix VI in [6]). Notice that (3.30) also represents the relation between I and B_m along a given field line or, in general, on a given L-shell in a dipole field.

In non-dipole trapping fields, one still can use relationship (3.30) to assign an L-value to a particle of given I and B_m values (calculated with the pertinent magnetic field model through (2.33) and (2.37), respectively):

$$L = \left(\frac{B_E}{B_m}\right)^{1/3} f\left(\frac{I^3 B_m}{R_E^3 B_E}\right) \tag{3.31}$$

The function f is the cubic root of F. However, defined in this way, L will *not* be constant along a field line as in the pure dipole field case, and it will lose its geometric meaning as the radial distance to the equatorial point of its guiding field line. Still, it is a useful parameter in "quasi-dipole" fields like the magnetosphere out to about $L = 7$ during quiet times, giving intuitive information about the radial extension of a drift shell. It is certainly more convenient than the original invariants I or the energy-dependent J, both of which of course vary strongly along a field line. For near-equatorial particles ($\mu_0 \to 0$, $I \to 0$) an expansion of function f in powers of its argument can be used. Using (3.12) we obtain to first order

$$L \simeq \left(\frac{B_E}{B_0}\right)^{1/3} + \mu_0^2 \left[\frac{1}{R_E}\sqrt{\frac{B_0}{a_0}} - \frac{1}{3}\left(\frac{B_E}{B_0}\right)^{1/3}\right] \tag{3.32}$$

This shows that for a pure dipole field L is independent of μ_0 because in that case $a_0 = \partial^2 B/\partial s^2 = 9(B_0/R_E^2)(B_E/B_0)^{-2/3}$ and the bracket is zero. For a dipole-like field with additional internal and external distortions, the definition of L in (3.31) still gives as *zeroth* approximation the radial distance in Earth radii to the equatorial point of a pure dipole field line with the same equatorial magnetic intensity B_0.

Just as velocity maps are a useful graphic instrument (e.g., Fig. 2.12) for the representation of field-aligned properties of trapped particles, so are $B - L$ maps for the representation of stably trapped particle properties on a drift shell in dipole-like fields. This "$B - L$ space" has a physical region bounded by the equatorial field $B_0(L) = B_E/L^3$ and the Earth intersection field $B_L(L) = B_E\sqrt{4 - 3/L}$ (3.25), as shown in Fig. 3.15. Drift shells map into vertical lines between the two curves; a given $B - L$ ring (e.g., a mirror point trajectory) maps into a point.

The Earth's internal magnetic field is not that of a pure dipole. Within about 4 R_E, where contributions from external currents can be neglected during magnetospherically quiet times, the internal geomagnetic field is indeed mainly dipolar with very slowly time-varying perturbations caused by irregularities in the electric current flow in the Earth's core and effects of the irregular distribution of matter with varying magnetic properties in the mantle and rocky crust. Solid earth geophysicists express the internal magnetic field above the earth surface as deriving from a scalar magnetic potential $V_m(r, \theta, \phi)$ where r, θ, ϕ are the geographic spherical coordinates of the point in question:

$$V_m(r, \theta, \phi) = R_E \sum_{n=1} \sum_{m=0}^{m=n} \left(\frac{R_E}{r}\right)^{n+1} \left(g_n^m \cos m\phi + h_n^m \sin m\phi\right) P_n^m(\theta) \tag{3.33}$$

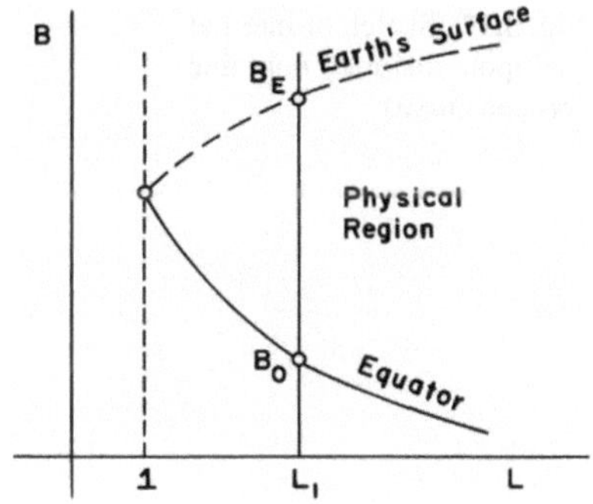

Fig. 3.15 The "good old" $B - L$ space and limiting curves defining the physically meaningful portion

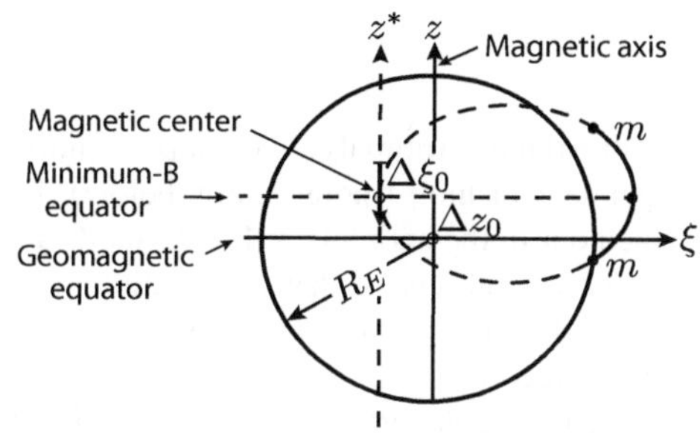

Fig. 3.16 Field line configuration of an off-center geomagnetic dipole

P_n^m are the associate Legendre functions, and the coefficients g_n^m and h_n^m are determined from data of the world network of magnetic observatories; they vary in time on a scale of decades to centuries (secular variation). The index n represents the multipole order, whose strength decreases with distance like $r^{-(n+1)}$. A more convenient and more physical form of representation of the internal field for space studies is obtained using the *geomagnetic coordinate system*, oriented in such a way that only one of the order $n = 1$ coefficients is non-zero ($g_1^0 \simeq 0.31$ Gauss; $g_1^1 = h_1^0 = h_1^1 \equiv 0$). In that system, the $n = 1$ dipolar axis is parallel to the polar axis. However, the dipole is not centered: the quadrupole terms with $n = 2$ and $m = 0, 1$ control the displacement of the main dipole away from the Earth's center. Since trapped particle drift shells are "frozen" into the magnetic field, their resulting geometric situation with respect to planet Earth is shown in Fig. 3.16 (exaggerated scale). In other words, the Earth sits off-center in the radiation belts (solid earth geophysicists would prefer a reverse statement!) with the result that it takes a bite out of them in a geographically limited region in the South Atlantic where mirror points come closest to the ionosphere, and loss cones widen,[5] and trapped particles may precipitate into the ionosphere (the area is called the

[5]This is precisely how James Van Allen and his group [10] discovered the radiation belt with the Explorer I and III satellites: a cosmic ray counter blacked out every time the satellites crossed the South Atlantic region. A careful analysis of the orbital points of saturation and recovery of the instrument and of the respective local magnetic field intensities led the scientists to the conclusion that the saturation had to be caused by a high flux of energetic particles trapped in the Earth's magnetic field (initially it was not clear whether the radiation was natural or consisting of trapped

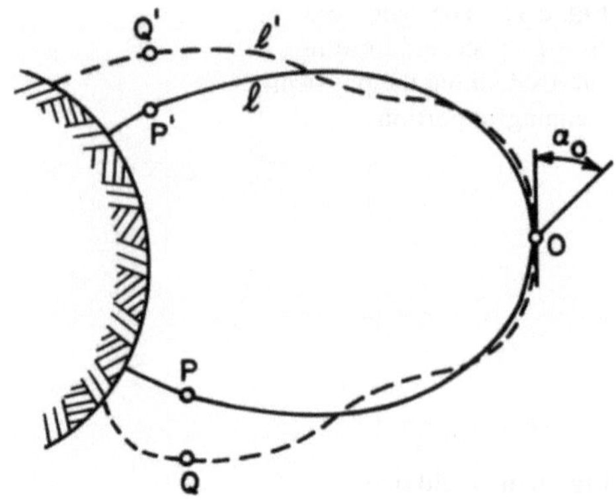

Fig. 3.17 Sketch of internal multipole-distorted field line (broken curve)

South Atlantic Anomaly). The upper limiting curve in Fig. 3.15 dips down in longitude-dependent fashion (more details below). Electrons, which drift from west to east, are wiped out in the anomaly at low L-shells and smaller pitch angles, but the flux recovers rapidly several tens of degrees longitude east of that area due to pitch angle scattering, which fills the emptied loss cones (in the early times this was called the "windshield wiper effect").

The multipole terms give rise to field distortions which decrease with distance faster than the dipole field. Thus the field distortion caused by them is much greater near the surface than further out in space. Let us imagine that we can switch these higher order multipoles on and off at will. If we start with a pure dipole field, the field line going through the equatorial point O is ℓ (Fig. 3.17). Now we turn on all higher multipoles. The field line through the same point O (which may no longer may be a minimum-B point) may look like the broken curve ℓ'. It will differ only very little from line ℓ near the equator, but may depart considerably from it near the earth.

A particle injected with a given pitch angle α_0 at O along the dipole field line ℓ will mirror at points P, P′ where the field intensity is $B_m = B_o/\sin^2\alpha_0$ (2.33). Its L-value will be given by (3.22). The same particle, injected with the same pitch angle along the field line ℓ' with all multipoles turned on, will mirror at the pair Q, Q′, different from P, P′ but where the field intensity is practically the same (different only to the extent that the *equatorial* field intensity B_0 differs a tiny bit from the original dipole value—see relation (2.33)). Will its I value be very different? Since the mirror points Q, Q′ are at positions different from P, P′, one may expect the I-value to be substantially different, too. Yet numerical calculations for the real geomagnetic field show that this is not the case. The reason for this near-equality is that the integrand $\sqrt{1 - B(s)/B_m}$ in (2.37) mainly contributes in the equatorial region where the multipole effect is strongly attenuated; toward the mirror points, where the multipole field distortion becomes more and more pronounced, the integrand decreases toward zero. The situation is quite different with the bounce period τ_b (2.34) or the half-bounce path length S_b (2.35) whose

beta decay electrons from fission products of secret high altitude explosions.) At nearly the same time, Sergei Vernov and his groupt at Moscow State University [11] saw the same effect happening to cosmic ray detectors on Sputnik satellites, but initially he attributed it to equipment failure.

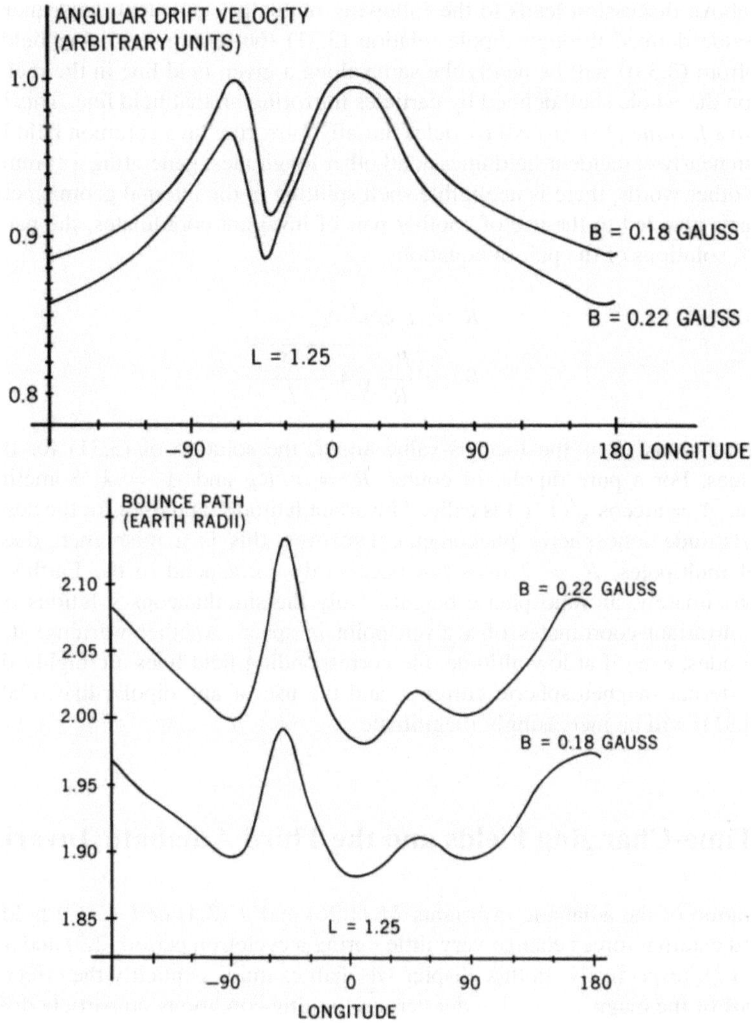

Fig. 3.18 Angular drift velocity (in arbitrary units) and bounce path S_b for particles at $L = 1.25$ and mirroring at $B_m = 0.18$ and 0.22 Gauss, respectively [12] (To obtain drift velocity in degrees/sec, multiply by factor given for Fig. 3.11)

integrands contain the same function but in the denominator. Its main influence now occurs in the region near the mirror points, causing bounce period and bounce path to differ appreciably from the dipole case (remember that a trapped particle always spends more time of its bounce oscillation near its mirror points than around the equatorial point!). Figure 3.18 shows the variation with longitude due to internal geomagnetic field multipoles of the bounce-average drift velocity and the bounce path, for particles at $L = 1.25$ mirroring near the Earth surface at $B_m = 0.18$ and 0.22 Gauss.

The above discussion leads to the following mutually equivalent consequences: (i) The L-value defined through dipole relation (3.31) (but computed using field values derived from (3.33)) will be nearly the same along a given field line in the real field as well as on the whole shell defined by particles mirroring on that field line. This is called *McIlwain's L value* [13]. (ii) All particles initially mirroring on a common field line will mirror on nearly coincident field lines at all other longitudes, generating a common drift shell; in other words, there is negligible shell splitting in the internal geomagnetic field. These facts also led to the use of another pair of invariant coordinates, the parameters R and Λ, solutions of the pair of equations

$$\begin{aligned} R &= L\cos^2\Lambda \\ B &= \frac{B_E}{R^3}\sqrt{4-\frac{3R}{L}} \end{aligned} \tag{3.34}$$

In these relations, B is the local B-value and L the solution of (3.31) for the local B, I values. For a pure dipole, of course $R = r/R_E$ and $\Lambda = \lambda$. Sometimes the parameter $\Lambda = \arccos\sqrt{(1/L)}$ is called "invariant latitude" and used for the description of high-latitude ionospheric phenomena. However, this is a misnomer: due to the effect of multipoles, $R = 1$ does *not* necessarily correspond to the Earth's surface (or, approximately, an ionospheric height). Only the simultaneous solutions of (3.34) are true invariant coordinates of a given point in space. Another warning: if used at high latitudes, even if at low altitude, the corresponding field lines are highly distorted by the external magnetospheric currents, and the use of any dipolar-like relationship (as in (3.31)) will be increasingly illegitimate.

3.5 Time-Changing Fields and the Third Adiabatic Invariant

Conservation of the adiabatic invariants M (1.26) and J (3.4) or I (2.37) hold only if fields and external forces change very little during a cyclotron period (2.2) and a bounce period (3.8), respectively. In this chapter we shall examine explicitly the effect of *time variations* of the magnetic field under certain limiting conditions on particle drift shells and energy.

Let us imagine a trapping magnetic field configuration whose time variations can be turned on and off, at our own will. We first impose the restriction of dealing only with particles which during static conditions are on closed shells (i.e., stably trapped whenever time variations are turned off). In this situation, the drift motion represents the third periodicity of the general equation of motion (1.1).

Quite often one is not interested in the actual drift phase of a particle; in such a case the drift motion may be averaged out. One can still investigate the configuration and time-change of the "bare" drift shell without worrying about where on that drift shell the particle is located at any given time. This is physically analogous to averaging out the cyclotron phase and working with the guiding center, or to averaging out the bounce motion and working with the guiding field line, without specifying the cyclotron or bounce phases, respectively. The quantity

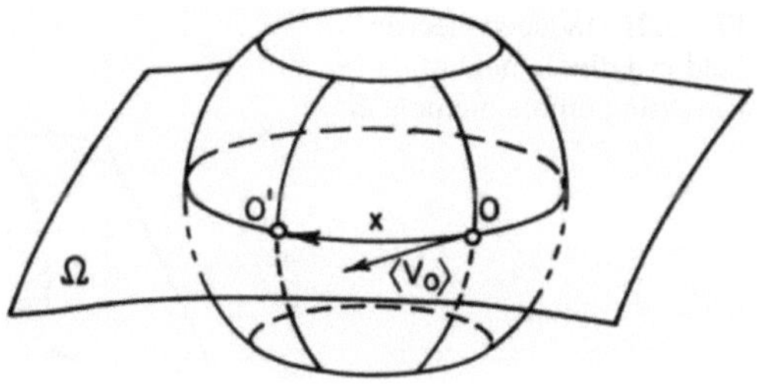

Fig. 3.19 Parameters of the drift shell of a positive particle in the earth's field

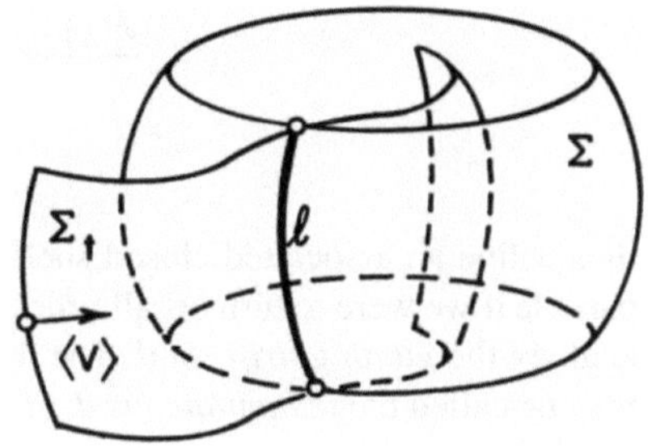

Fig. 3.20 Illustrating the concept of instantaneous guiding drift shell in a time-dependent field

$$\tau_d = \oint \frac{d\ell}{\langle V_0 \rangle} \tag{3.35}$$

is the *drift period* ($\langle V_0 \rangle$ is the bounce-average drift velocity (3.7)). The integral is computed along a closed line on the drift shell (Fig. 3.19) under *static* conditions (with all time variations turned off). In (3.35) $d\ell$ is the element of arc length of the intersection of the drift shell with the reference surface Ω (usually the minimum-B surface). Whereas $\langle V_0 \rangle$ and the integration path depend on the reference surface, τ_d does not. Under the conditions (2.2) and (3.8) for conservation of M and J, we have $\tau_c \ll \tau_b \ll \tau_d$; any field time variation $\dot{B}$ that fulfills the condition

$$\tau_d \ll \frac{B}{\dot{B}} \tag{3.36}$$

may be called a "fully adiabatic" change. The other restrictive condition which we impose for our discussion in this section is that no non-electromagnetic forces act on the particle.

We now switch on a time variation of the magnetic field. A force will appear, given by the *induced electric field* associated with the time-dependent magnetic field. This induced electric field E_{ind} causes an additional drift (1.34) which drives the particle out of its initial shell. During the interval of transient conditions the particle's drift shell may not be closed. Consider such a transient drift shell Σ_t as shown in Fig. 3.20. At a given time t the particle's guiding field line is $\boldsymbol{\ell}$. If at that instant we turn off the time variation and "freeze" the magnetic field configuration, we may again have the particle following a static closed drift shell Σ. For each time and position on the transient shell we can

Fig. 3.21 Induced electric field in a dipole field of increasing dipole moment k_0

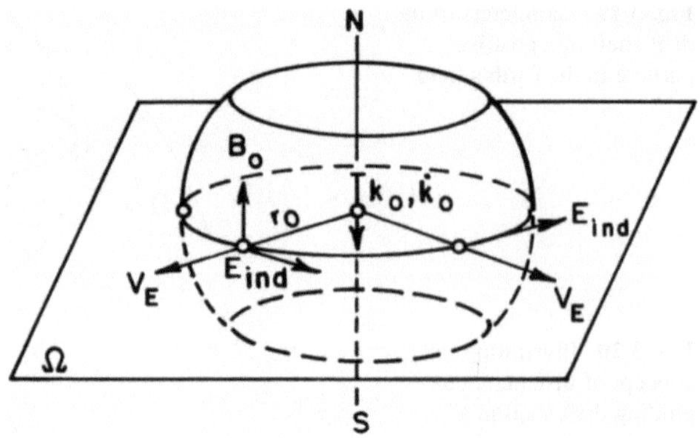

thus define an associated closed shell, which is the one that would be generated by the particle if we were to turn off all time variations at that instant. We call this static shell by analogy the *guiding drift shell*. The drift period of the particle on the guiding drift shell may be called the *instantaneous drift period* of the particle. During static conditions the guiding drift shell represents the actual drift shell of a particle, just as a guiding field line represents the actual portion of a field line on which a particle is momentarily bouncing. When the field changes, a particle is transferred from one guiding drift shell to another by the drift caused by the induced electric field.

For instance, if we have a dipole field and gradually increase the dipole moment k_0, we will induce an azimuthal electric field of the direction as shown in Fig. 3.21 (using Faraday's law, $E_{ind} = (dk_0/dt)\, r^{-2}$). This electric field produces an *outward* drift (for positive *and* negative charges) and the guiding drift shell will inflate. Likewise, a decrease of the dipole moment causes guiding drift shells to contract. A change in drift shell is, of course, accompanied by a change in particle energy. In order to compute this change, we envisage Eq. (2.26) and average it over a bounce period (in absence of external forces):

$$\left\langle \frac{dT}{dt} \right\rangle_b = M \left\langle \frac{\partial B}{\partial t} \right\rangle_b + q \langle \boldsymbol{V} \cdot \boldsymbol{E}_{ind} \rangle_b$$

As to the second term $\langle \boldsymbol{V} \cdot \boldsymbol{E}_{ind} \rangle_b$, we first notice that it is equal to $\langle \boldsymbol{V}_{GC} \cdot \boldsymbol{E}_{ind} \rangle_b$, in view of the fact that there are no external forces other than $q\boldsymbol{E}_{ind}$ and that the drift component $(\boldsymbol{E}_{ind} \times B)/B^2$ caused by $\boldsymbol{E}_{ind}$ is perpendicular to it ($\boldsymbol{V}_{GC}$ is the gradient-curvature drift (2.17)). It can be shown (on similar lines as for (3.6)) that

$$\langle \boldsymbol{V}_{GC} \cdot \boldsymbol{E}_{ind} \rangle_b = \langle \boldsymbol{V}_{GC0} \rangle_b \cdot \boldsymbol{E}_{ind0}$$

The subindex 0 refers to the minimum-B point of the guiding field line. We thus have, finally,

$$\left\langle \frac{dT}{dt} \right\rangle_b = M \left\langle \frac{\partial B}{\partial t} \right\rangle_b + q \langle \boldsymbol{V}_{GC0} \rangle_b \cdot \boldsymbol{E}_{ind0} \tag{3.37}$$

The first term represents what has been called the *gyro-betatron* acceleration effect (see also Sect. 2.2), because it gives the power delivered by the induced electric field force to the particle in its cyclotron motion mode. The second term represents the *drift-betatron*, i.e., the power delivered by the induced electric field to the particle's gradient-B drift mode.

In order to find the actual drift shell of a particle in a time-dependent magnetic field, we have to integrate the bounce-average total drift velocity (3.7), in which $\boldsymbol{E}_0 = \boldsymbol{E}_{ind}$, and make use of the conservation of M and $K = I\sqrt{B_m}$. The main practical problem lies with the determination of $\boldsymbol{E}_{ind}$, which has to be computed numerically. Assuming zero field-aligned electric fields and that the magnetic field lines are rooted into the ionosphere (frozen-in field lines, Sect. 5.5), one reasonable way is to (i) trace the field line from the minimum-B surface point $\boldsymbol{P}$ (at which $\boldsymbol{E}_{ind}$ is wanted) to the ionosphere; (ii) let the external magnetospheric field change in time Δt; (iii) re-trace back from the fixed ionospheric point to the new minimum-B point $\boldsymbol{P}' = \boldsymbol{P} + \boldsymbol{\Delta r}$; (iv) take the vector $\boldsymbol{\Delta r}/\Delta t\, B_0$ as a good approximation of $\boldsymbol{E}_{ind}$. As examples, consider two extreme cases.

(i) If the time variation is *fast* with respect to the drift period (but still slow compared to the bounce period, $\tau_b \ll B/\dot{B} \ll \tau_d$), the induced electric field drift will dominate over the gradient-curvature drift, and the particle will be driven mainly by the former during the transient interval. The drift velocity will be that of the guiding field line or, which is equivalent, the particle's guiding field line will be "frozen" into the changing magnetic field configuration and move with the latter. The acceleration (or deceleration) is then of "gyro-betatron" type (first term of (3.37)). If the time variation persists long enough, the particle may acquire sufficient energy to let the gradient-curvature drift take over and move it away from the initial moving field line. As a matter of fact, this represents a fundamental acceleration mechanism of radiation belt particles during geomagnetic substorms during which field lines are being "pulled from the tail towards the earth" in the midnight meridian region. Low energy particles are thus carried with these field lines by the induced electric field drift and accelerated by the gyro-betatron until the gradient-curvature drift takes over, ejecting them from their guiding field line and the acceleration region, eventually placing them on closed shells (stably trapped orbits). For the same reason, sudden asymmetric compressions of the magnetosphere caused by brusque variations of the dynamic pressure of the solar wind, can cause acceleration of particles found on the dayside and their placement into trapped orbits (see later). It should also be clear that, unless there is azimuthal symmetry both in drift shells and $\partial \boldsymbol{B}/\partial t$, particles belonging to the same drift shell but with different drift phases (e.g., longitude or arc position of their equatorial crossings) will in general not end up on a common drift shell as the time variation goes on—a fact that can lead to radial diffusion (see Chap. 4).

(ii) The other extreme case is given by time variations that are fully adiabatic, i.e., *very slow* compared to a drift period. In such a case a *third conservation theorem* applies, which can be conveniently used in numerical computations: The magnetic flux encompassed by the guiding drift shell of a particle remains constant, provided condition (3.36) is fulfilled:

$$\Phi = \oint \boldsymbol{A}_0 \cdot d\boldsymbol{\ell} = \text{const.} \tag{3.38}$$

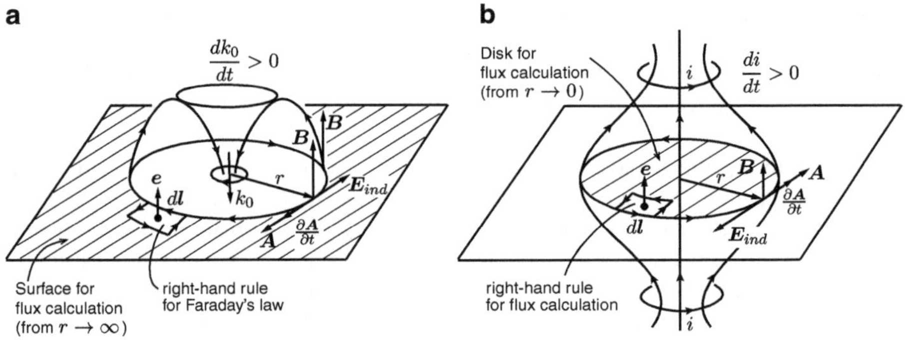

Fig. 3.22 Definition of the third invariant (flux Φ) for a dipole-like (**a**) and mirror machine-like (**b**) closed drift shell. In (**a**), because of the singularity at the origin, the flux Φ must be calculated over the surface lying outside the drift shell

A_0 is the magnetic vector potential and the integral is carried out along a curve which lies in the (time-frozen) guiding drift shell of the particle (e.g., intersection with the minimum-B surface; see Fig. 3.22). It is important to remind the reader that in a static situation in absence of any external forces, a drift shell is independent of the particle's kinetic energy and fully determined by just two invariants M (or B_m) and I (e.g., relations (3.9) and (3.10)). Φ is *not* an independent parameter under these conditions. For a time-dependent magnetic field, B_m and I are no longer invariant, and a shell must be determined by the general invariants M, J (or K in absence of field-aligned electric fields) and Φ, as we shall describe below.

The demonstration of the conservation theorem for Φ is lengthy [1, 14].[6] In Appendix A.3 we give a simplified demonstration limited to equatorial (90°) pitch angle particles. It is very important to emphasize that Φ is defined and computed for the *guiding* drift shell (i.e., holding the field constant and finding out what closed shell would be generated by the particle), and *not* for the actual drift shell of the particle during transient conditions. However, in view of the condition of slow changes (3.36), the actual drift shell will differ only very little from the instantaneous guiding drift shell. Indeed, we may picture this situation as having the particle drifting many times around a closed shell while the latter is slowly changing its shape. Note that if (3.36) holds for a given class of particles, because of the energy-dependence of the gradient-curvature drift this may not be true for other particles on the same initial shell. For a given scale $B/\dot{B}$ of time variation, the lower the particle energy, the more likely it is that condition (3.36) will be violated. On the other hand, particles of the same kind but different drift phases starting on a common drift shell may end up on different drift shells as the time-variation goes on. In Appendix A.3 we show under what conditions these particles do reassemble

[6] Or we may just refer again to analytical mechanics (see footnote on page 10) and remark that in the canonical path integral for the cyclic drift motion $J_d = \oint (\boldsymbol{p} + q\boldsymbol{A}) \cdot \boldsymbol{dl}$, the first term is zero to first order (the average $\boldsymbol{p}$ vector is just $m\boldsymbol{V}_D$), so that what remains is $J_d = \oint q\boldsymbol{A} \cdot \boldsymbol{dl} = q\Phi =$ const.!

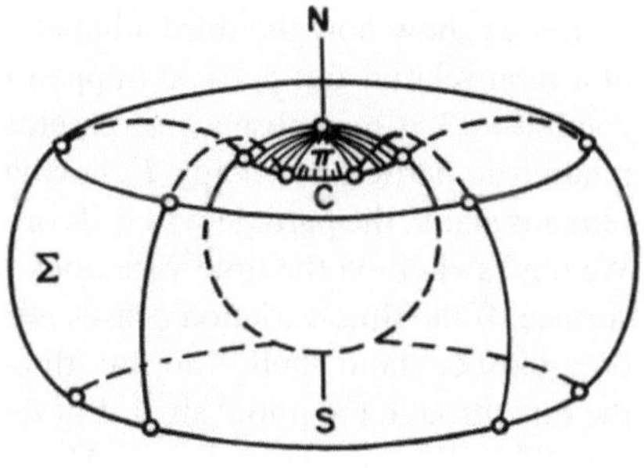

Fig. 3.23 Projection of a drift shell Σ onto the polar cap (π) along its magnetic field lines

on one common drift shell (integer multiples of τ_d!). In those cases the conservation of Φ is true only in *drift-average* terms. This is equivalent to what happens in shell splitting (Sect. 3.3): particles with different pitch angles on a common initial guiding field line will all pass through that same initial field line at integer multiples of drift time.

For a dipole field the magnetic flux encompassed by a particle shell defined by the parameter L (3.22) (Fig. 3.22) is given in absolute value by

$$\Phi = \frac{2\pi k_0}{r} = \frac{2\pi B_E {R_E}^2}{L} = \frac{1.953}{L} \ \text{Gauss} R_E^2 \tag{3.39}$$

For other magnetized planets, the corresponding value of B_E must be used. It is important to note that in this case and all dipole-like field geometries, Φ is equal to the flux $\int \boldsymbol{B} \cdot d\boldsymbol{S}$ integrated over the portion of the equatorial surface *outside* of the intersection O of the shell with Ω. This obliges us to be careful with the sense of integration of the vector potential in the determination of the flux (3.38).

For a more complex field geometry the integral (3.38) has to be carried out numerically. This requires knowledge of the vector potential $\boldsymbol{A}$. A more practical way for the closed field lines in the Earth's magnetosphere is to find the intersection C of a series of shell field lines with the earth's surface (Fig. 3.23), and to numerically compute the flux

$$\Phi = \int_{\pi} \boldsymbol{B} \cdot d\boldsymbol{S}$$

over the polar cap Π, using the earth's known *surface* field B_s in that cap. For most magnetospheric models it will be sufficient to use the dipole approximation for B_s. Calling $\lambda_C(\phi)$ the dipole latitude of the intersection C at a given longitude ϕ, the shell flux will be, in absolute value (using (3.24)):

$$\Phi \cong 2B_E {R_E}^2 \int_0^{2\pi} d\phi \int_{\lambda_C(\phi)}^{\pi/2} \cos\lambda \sin\lambda d\lambda = B_E {R_E}^2 \int_0^{2\pi} \cos^2[\lambda_C(\phi)] d\phi \tag{3.40}$$

A convenient "recipe" to evaluate numerically the third invariant in the outer magnetosphere is therefore: (i) Trace the particle's guiding shell by methods given in Sect. 3.2; (ii) Find the intersection dipole latitude $\lambda_C(\phi)$ as a function of longitude for a set of shell field lines tracing them down to the earth's surface; (iii) Integrate numerically (3.40).

Let us show how the third adiabatic invariant can now be used to determine the fate of a (non-relativistic) particle trapped in a time-dependent field subject to the adiabatic condition (3.36), in absence of external forces. We start with a static field and a non-relativistic particle of energy T, having invariant parameters I, B_m. As long as the field remains static, the particle will drift on a closed shell of flux Φ given by (3.38) or (3.40). We now switch on the time variations. The particle will drift away from the initial shell surface. If the time variation ceases after an interval Δt, the particle will again be found on a closed, static shell. The flux through the shell in this final state must be equal to the flux through the initial shell. But the energy of the particle T^* and its parameters I^* and B_m^* will have changed—yet $M = T^*/B_m^*$ and $K = I^*\sqrt{B_m^*}$ must be the same as in the initial state. Therefore, to determine the final shell, one proceeds as follows: (i) pick a (reasonable) pair of values I^* and B_m^* that satisfy $I^*\sqrt{B_m^*} = I\sqrt{B_m}$; (ii) trace the shell in the final field configuration for these two values, as explained on page 63; (iii) find the flux Φ^* (3.40) through this shell; (iv) iterate steps (ii) and (iii), changing I^* and B_m^* as many times as necessary until the final Φ^*-value is equal to the initial one ($\pm$ prefixed error); (v) compute the particle's final energy from $T^* = (T/B_m)B_m^*$.

Adiabatic time changes satisfying the condition (3.36) are reversible: when the field configuration comes back to the initial state, all particles come back to their initial drift shells. This follows directly from the conservation of Φ. For non-adiabatic time variations this is no longer true, even if M and J are still conserved. It is important to note that *in absence of external forces* shell behavior in an adiabatic time-dependent field is *independent of particle energy, mass and charge* (as happens in the static case). Theorem (3.38) holds even if external forces are acting. Moreover, it holds when B is static but the external forces are adiabatically time-dependent. One final remark on condition (3.36): if the time-dependent magnetic field is azimuthally symmetric *at all times* (i.e., symmetric field with symmetric perturbations), Φ happens to be conserved even if the adiabatic condition is *not* fulfilled.

We are now in the position to create an ideal, truly invariant *reference representation* of particles found stably trapped in any non-symmetric magnetospheric field configuration. For that purpose, we need a sophisticated numerical model of the magnetic field, with parameters that can be conveniently adjusted to reproduce quantitatively the trapping magnetic field configuration of the moment (e.g., based on some real-time magnetic field measurements in strategic regions of the magnetosphere). Suppose we have such a sophisticated model on hand. For a given particle of energy T, pitch angle α and position $\boldsymbol{r}$, the model can be used to determine numerically the local magnetic field intensity B and the particle's invariant parameters M, K and, through shell tracing, Φ. Now we imagine turning off very slowly all external field sources (e.g., boundary and neutral sheet currents) and, if we are close to the Earth's surface, also the internal multipoles, leaving just the field of a central pure dipole of moment $k_0 = B_E R_E^3$. The particle, while conserving the values of its adiabatic invariants, will end up in a beautifully symmetric dipole drift shell defined by what we can call the *adiabatic reference parameters* and designate as "L-star" L^* and "B-star" B_m^*, respectively.[7] The particle kinetic energy, of course, will also be different, T^*.

[7]In the literature L^* has sometimes been called the "Roederer L-value" to distinguish it from the "McIlwain L-value" (3.30).

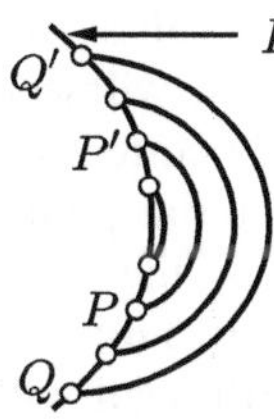

Fig. 3.24 Mirror points on different field lines for which trapped particles have the *same* L^* value

The adiabatic conservation theorems allow us to calculate the values of these reference parameters. The determination of L^* is straightforward, based on relation (3.39):

$$L^* = \frac{2\pi B_E {R_E}^2}{\Phi} \tag{3.41}$$

Once we have L^*, we can use dipole relation (3.31) to determine B_m^* by replacing on the right side the I^*-value by $I^* = K(B_m^*)^{-1/2}$ and solving the functional relationship for B_m^*. The particle energy is then obtained through the conservation of M: $T^* = T(B_m^*/B_m)$ (relativistically, we would obtain for the momenta $p^{*2} = p^2(B_m^*/B_m)$). If we are using the transformation to a stationary reference dipole field for standard flux mapping purposes, also the *flux values must be transformed*; this will be addressed in the next chapter. Finally, it is important to be well aware of the fact that a surface of constant L^* (see Fig. 3.24) represents neither a particle shell nor a collection of field lines. It simply gives the locus of all mirror points $P, P' \ldots Q, Q' \ldots$ of particles whose drift shells have the same Φ-value. Transforming adiabatically into a pure dipole field, all mirror points shown in this figure will assemble on one single dipole field line and shell surface (with different energies though, if initially they all had the same). Conversely, the L^* value (or Φ) will vary along a given field line in the real field because particles mirroring at different points of that field line generate different drift shells with different Φ values.

We can figure out the differences in Φ or L^* values. Let us use the same approximations for the field and other field-related quantities for near-equatorial particles as in Sects. 1.6, 3.2 and 3.3 (particularly (3.19) and (3.20)), to obtain following approximate expression for the L^* value of a particle injected at the equatorial point r_0, ϕ_0 with a pitch angle close to 90° ($\mu_0 \ll 1$):

$$L^*(r_0, \phi_0, \mu_0) \simeq \frac{r_0}{R_E}\left(1 + \frac{1}{2}\frac{b_1}{B_E}\left(\frac{r_0}{R_E}\right)^3 - \frac{1}{3}\frac{b_2}{B_E}\left(\frac{r_0}{R_E}\right)^4\left(1 - \frac{43}{18}\mu_0^2\right)\cos\phi_0\right) \tag{3.42}$$

Note the effects of magnetospheric compression (b_1 term) and day-night asymmetry (b_2 term).

The quantity

$$\sigma_A = -\frac{1}{3}\frac{b_2}{B_E}\left(\frac{r_0}{R_E}\right)^5\left(1 - \frac{43}{18}\mu_0^2\right)\cos\phi_0 \tag{3.43}$$

may be called the shell asymmetry parameter. According to (1.50) any changes of the stand-off distance R_s will lead to changes in the radial parameter L^*. It is important to interpret Eq. (3.42) correctly: it gives the approximate L^* value of a particle injected at an *initial* equatorial point with an *initial* pitch angle under magnetospheric field conditions given by the constants b_1 and b_2. If these constants change *adiabatically* (slowly compared to a drift period), the value of L^* will remain constant (conservation of Φ!) but the drift shell will change. To obtain the new drift trace one has to solve (3.43) for $r_0 = r_0(\phi_0)$ and constant L^* given by the initial conditions. For fast changes of b_1 and b_2, the drift-average of the change in L^* (only the constant b_2 will contribute) will give a measure of the effect on third invariant changes (see next chapter).

The dimensionless quantity

$$\sigma_S = \frac{\partial^2 L^*}{\partial \mu^2} = \frac{43}{54}\frac{b_2}{B_E} L^{*5} \cos\phi_0 \tag{3.44}$$

can be taken as a quantitative measure of shell splitting (second derivative, because the first derivative is zero on the equator). Only in the case of shell degeneracy, i.e., for azimuthally symmetric fields, will L^* and Φ be constant along a field line and $\sigma = 0$. Notice that $\sigma_S < 0$ on the day side ($\cos\mu < 0$) and positive on the night side. In other words, particles mirroring at higher latitude (i.e., with smaller equatorial pitch angles) on the noon/midnight field line, have smaller/larger L^*-values (larger/smaller third invariant Φ). This is directly related to the inward/outward reversal of shell splitting for noon/midnight field line particles (see Figs. 3.9 and 3.10). The drift-average of σ_S^2 is a good measure of the influence of the general state of the magnetosphere (represented by the parameter R_s) and the asymmetry of drift shells and its relation to drift shell splitting. As such it plays a role in radiation belt diffusion. Taking into account (1.50) we can establish the following proportionality (the actual factors of proportionality are magnetospheric model dependent):

$$\langle \sigma_S^2 \rangle \sim R_s^{-8} L^{10} \tag{3.45}$$

Notice the strong dependence on the stand-off distance of the magnetospheric boundary and the even stronger dependence on the radial shell parameter L.

Turning back to the general case of off-equatorial particles, one important remark. We often talk about the L^*-value of this or that point $\boldsymbol{r}$ in real 3-D space. One has to have clearly in one's mind what is meant by that, because, in general, the drift shell of a particle will depend on its energy. Can we even talk of the L^*-value "of a point" (Fig. 3.24) in such a case? *Yes we can*—provided we define it clearly: it is the L^*-value of a particle *mirroring at that point* when the *magnetic field is held constant in time*, and when *all electric fields have been turned off.*

Finally, Fig. 3.25 gives an example for particles with a series of different pitch angles detected at several different local times in synchronous orbit ($r = 6.6\,R_E$), using a magnetospheric field model for quiet time [15]. The picture clearly shows the portions of invariant reference B^*-L^* space scanned by such detectors. In a pure dipole field, each detector would just "see" one point in B, L space; there would be no local-time

Fig. 3.25 Coordinates in invariant $B^* - L^*$ space (see text) of the particles registered with a directional detector in synchronous orbit

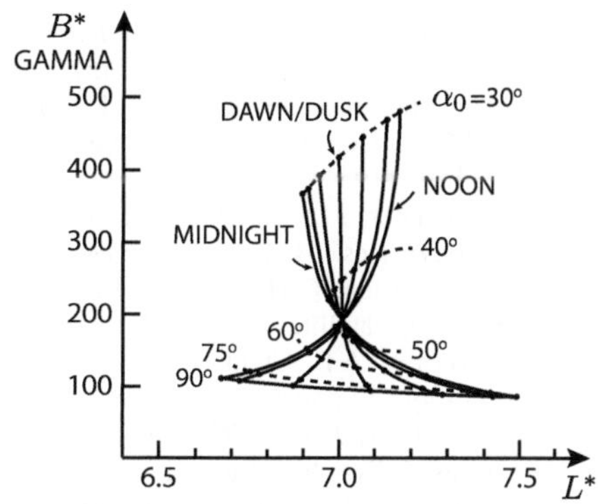

Fig. 3.26 Same as Fig. 3.25, in McIlwain's $B - L$ space

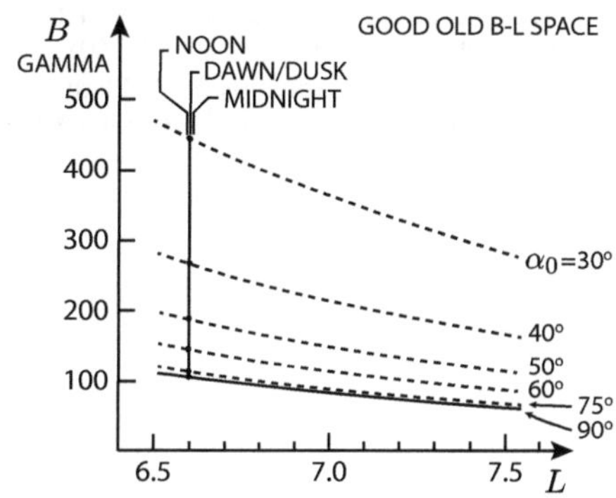

effect, and the region sampled would just be a straight vertical line (Fig. 3.26). For increased magnetospheric compression, the pattern in the figure spreads out in L^* coverage. Notice in Fig. 3.25 that for pitch angles around 45°–50°, the whole local-time asymmetry disappears and its effect reverses. This is related to what we have mentioned on page 72 in connection to shell splitting and its reversal at what we have called a "magic" pitch angle, because it barely changes as a function of specific parameters governing the external current contributions in the magnetospheric model. We shall come back to this invariant mapping procedure in the next chapter.

References

1. T.G. Northrop, *The Adiabatic Motion of Charged Particles* (Interscience Publishers, New York, 1963)
2. D.P. Stern, Euler potentials. Am. J. Phys. **38**, 494–501 (1970)
3. K. Min, J. Bortnik, J. Lee, A novel technique for rapid L* calculation using UBK coordinates. J. Geophys. Res. **118**, 192–197 (2013)
4. E.C. Stone, The physical significance and application of L, B_0, and R_0 to geomagnetically trapped particles. J. Geophys. Res. **68**, 4157–4166 (1963)
5. D.J. Williams, G.D. Mead, Nightside magnetosphere configuration as obtained from trapped electrons at 1100 kilometers. J. Geophys. Res. **70**, 3017–3029 (1965)

6. J.G. Roederer, *Dynamics of Geomagnetically Trapped Radiation* (Springer, New York, 1970)
7. K. Min, J. Bortnik, J. Lee, A novel technique for rapid L* calculation: algorithm and implementation. J. Geophys. Res. **118**, 1912–1921 (2013)
8. G.D. Mead, Deformation of the geomagnetic field by the solar wind. J. Geophys. Res. **69**, 1181–1195 (1964)
9. J.G. Roederer, M. Schulz, Effect of shell splitting on radial diffusion in the magnetosphere. J. Geophys. Res. **74**, 4117–4122 (1969)
10. J.A. Van Allen, G.H. Ludwig, E.C. Ray, C.E. McIlwain, Observations of high intensity radiation by satellites 1958 Alpha and Gamma. Jet Propuls. **28**, 588–592 (1958)
11. S.N. Vernov, A.E. Chudakov, E.V. Gorchakov, J.L. Logachev, P.V. Vakulov, Study of the cosmic-ray soft component by the 3rd Soviet Earth Satellite. Planet. Space Sci. **1**, 86 (1959)
12. J.G. Roederer, J.A. Welch, J.V. Herod, Longitude dependence of geomagnetically trapped electrons. J. Geophys. Res. **72**, 4431–4447 (1967)
13. C.E. McIlwan, Magnetic coordinates. Space Sci. Rev. **5**, 585–598 (1966)
14. T.G. Northrop, E. Teller, Stability of the adiabatic motion of charged particles in the Earth's field. Phys. Rev. **117**, 215–225 (1960)
15. J.G. Roederer, Geomagnetic field distortions and their effects on radiation belt particles. Rev. Geophys. Space Phys. **10**, 599–630 (1972)

Chapter 4
Particle Fluxes, Distribution Functions and Violation of Invariants

4.1 Particle Fluxes and Pitch Angle Distributions

So far we have been concerned with individual particle or guiding center motion in given magnetic and electric fields. In the real magnetosphere, however, we always deal with large *ensembles* of particles of different species, different energies, velocities and spatial densities. In this chapter we will describe how particle *distributions* are represented and how they are handled within the framework of adiabatic theory, i.e., when particles exhibit a consistent "micro-behavior" of cyclotron motion, a "meso-behavior" of bounce motion, and a "macro-behavior" of drifts. We will come across two types of distributions, the actual or "kinetic" distributions of the original particles, and the distributions of their guiding centers, in which information on their cyclotron motion has been averaged out and replaced by a magnetic moment (Sect. 1.2).

Suppose we have an ideal particle detector of sensitive area $\delta A_\perp$ that registers the incidence of particles of a given class (charge, mass) traveling in a direction $\boldsymbol{u}$ perpendicular to the area, lying within a very narrow solid angle cone $\delta\Omega$ (Fig. 4.1), and whose kinetic energy falls into the interval T and $T + \delta T$ (non-relativistic energies). If δN is the number of such particles arriving in the time interval δt, we can write

$$\delta N = j\, \delta A_\perp \delta\Omega \delta T \delta t \tag{4.1}$$

The constant of proportionality j is called the *directional differential flux* of the particles; in general, it will be a function of position, time, energy and direction. It represents the average number of particles per unit area perpendicular to their direction of movement, per unit of time, energy and solid angle; in magnetospheric physics its units are usually chosen as $m^{-2}s^{-1}kev^{-1}ster^{-1}$ (a steradian being the solid angle of a cone of aperture = 1 radian). The directional flux can be conferred

J.G. Roederer and H. Zhang, *Dynamics of Magnetically Trapped Particles*, Astrophysics and Space Science Library 403, DOI 10.1007/978-3-642-41530-2_4,

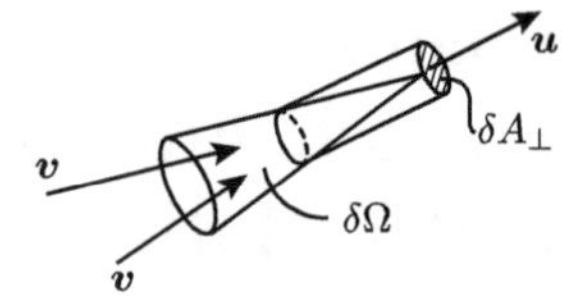

Fig. 4.1 Most relevant aspects of the geometry of a particle detector

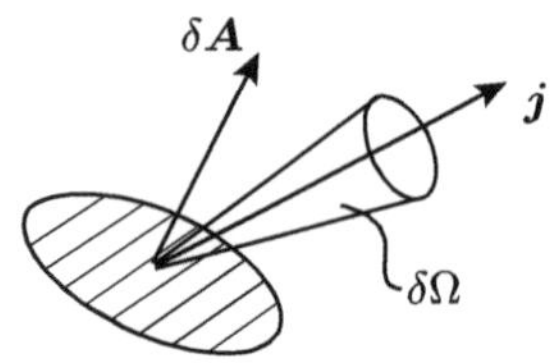

Fig. 4.2 Definition of directional flux $\boldsymbol{j}$

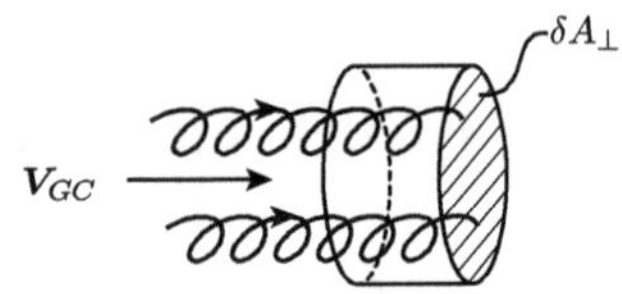

Fig. 4.3 Case of a "large" detector counting guiding centers

vectorship by setting $\boldsymbol{j} = j\boldsymbol{u}$, so that the number of particles traversing an element of area $\boldsymbol{\delta A}$ oriented in *any* direction $\boldsymbol{n}$ (Fig. 4.2) is

$$\delta N_A = j\boldsymbol{u} \cdot \boldsymbol{n}\delta A \delta\Omega \delta T \delta t = \boldsymbol{j}(\boldsymbol{r}, t, T, \boldsymbol{u}) \cdot \boldsymbol{\delta A}\delta\Omega\delta T\delta t \tag{4.2}$$

In all this we have assumed that the area $\delta A \ll \rho_C^2$. On the other extreme, if $\delta A \gg \rho_C^2$ the detector would be mainly counting guiding centers (Fig. 4.3) traveling in the direction $\boldsymbol{u}$. In experimental radiation belt physics, most detectors are of the first kind; their actual output is directional *counting rate*, which under ideal conditions of collimation and efficiency is proportional to the local flux $j(\boldsymbol{r}, t, T, \boldsymbol{u})$ in the direction $\boldsymbol{u}$.[1] Note that the directional flux vector $\boldsymbol{j}(\boldsymbol{r}, t, T, \boldsymbol{u})$ is not a traditional vector field: it represents a manifold, assigning a number to each point of space *and direction*. It is a local characteristic of the particle ensemble, now viewed as a macroscopic model fluid in which the original discontinuous stream of particles is replaced by a flowing continuum.

It is important to realize that in the adiabatic motion picture, a directional flux in the direction $\boldsymbol{u}$ does not necessarily imply that those particles actually come from the direction $-\boldsymbol{u}$ when viewed on a larger scale. For instance, consider the ensemble of 90° pitch angle particles in a homogeneous magnetic field as shown in Fig. 4.4a:

[1] An ideal differential detector gives *counting rate* $\delta N/\delta t$ as direct output, i.e., measures j, but real detectors have transmission functions that must be determined in careful calibration measurements, and they respond in *finite* intervals of energy ΔT, direction and solid angle $\Delta\Omega$. The conversion of instrument counts to directional particle fluxes is usually a difficult task for the experimentalist.

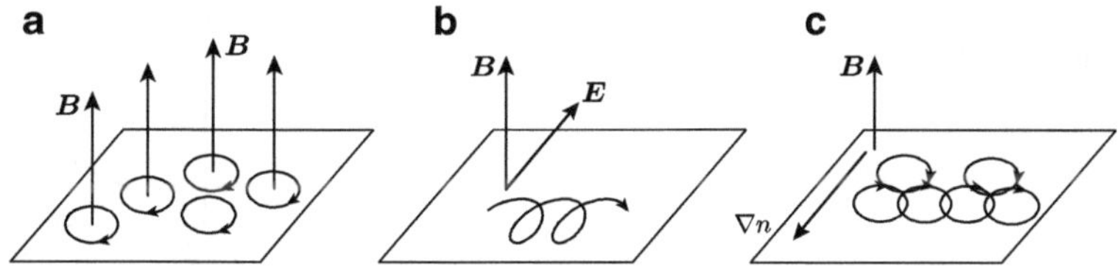

Fig. 4.4 Three cases of 90° pitch angle, monoenergetic particles—see text

an ideal directional detector would respond only when it is pointing perpendicularly to $\boldsymbol{B}$ for any azimuthal orientation φ (independent of φ, if the individual cyclotron phase angles are random). The detected particles, while locally coming from the direction $-\boldsymbol{u}$, are really just circling in situ around their guiding center. A large detector (with $\delta A > \rho_C^2$) in principle would not respond at all, because there is no flow of guiding centers. If now we impose an electric field $\boldsymbol{E} \perp \boldsymbol{B}$ (Fig. 4.4b), the particles will drift; a small detector will again respond only when oriented perpendicularly to $\boldsymbol{B}$, but the response will be φ-dependent because their energy will vary (1.35) and more particles will fall in from the direction of the drift (1.10). A large detector will register, too, because of the flow of guiding centers. Finally, if there is a gradient in particle density perpendicular to $\boldsymbol{B}$ (in absence of $\boldsymbol{E}$), there will be an anisotropic response in the plane perpendicular to the magnetic field, due to the effect depicted in Fig. 4.4c. In all of this, we have assumed that the detectors were at rest; two detectors that move with respect to each other will see different flux functions. Usually, a flux function is defined in the original frame of reference; in some cases it will be necessary to transform $\boldsymbol{j}$ into the local guiding center system.

In what follows, we shall ignore the space and time dependence of j, i.e., we shall concentrate on the flux of particles at one given point and time and just write $j = j(T, \boldsymbol{u})$. The T dependence of j is called the *differential energy spectrum* of the particles (traveling in direction $\boldsymbol{u}$ through the point $\boldsymbol{r}$ at time t); the dependence on the direction $\boldsymbol{u}$ is called the angular distribution. In addition, we have the following designations:

$$I(T, \boldsymbol{u}) = \int_0^T j(T', \boldsymbol{u})dT' \quad \text{integral spectrum flux} \tag{4.3}$$

$$F(T) = \int_{4\pi} j(T, \boldsymbol{u})d\Omega \quad \text{omnidirectional differential flux} \tag{4.4}$$

All these functions are, of course, also functions of $\boldsymbol{r}$ and t.

In radiation belt and magnetospheric plasma physics one usually uses the natural coordinate system (Appendix A.1) determined by the local magnetic vector field (measured, or as given by a numerical field model), and polar coordinates (Fig. 4.5)

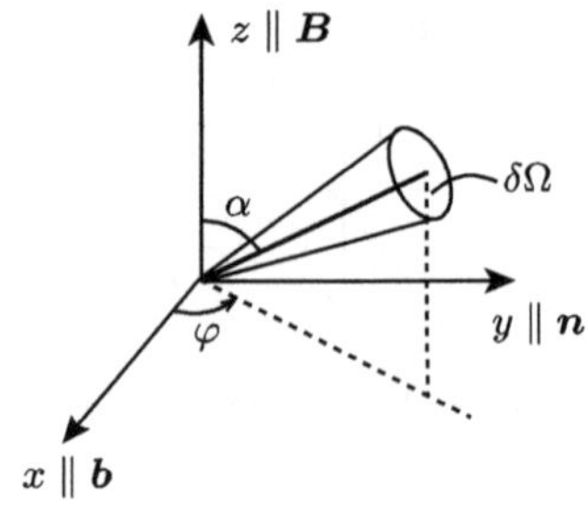

Fig. 4.5 Coordinates in the natural frame of reference

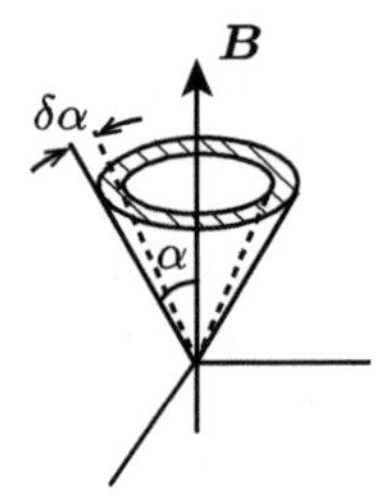

Fig. 4.6 Geometric parameters concerning the discussion of gyrotropic distributions

for the direction. If $\mu = \cos\alpha$, where α is the particle's pitch angle, $\delta\Omega = \delta\mu\delta\varphi$, we write

$$j = j(T, \mu, \varphi)$$

at a given point and instant of time. The unit directional vector $\boldsymbol{u}$ has components $(\sqrt{1-\mu^2}\cos\varphi; \sqrt{1-\mu^2}\sin\varphi; \mu)$. Note that j represents the number of particles traversing the surface element per unit time, unit area (perpendicular to $\boldsymbol{u}$) and unit energy. The vector $\boldsymbol{j}_\perp = \boldsymbol{j}(T, 0, \varphi)$, sometimes called "jay-perp", is the flux of particles mirroring at or near the point in question.

If $\partial j/\partial\varphi = 0$, the flux is called *gyrotropic* (random cyclotron phases). In such a case, one often integrates $(\int j d\varphi)\delta\mu = 2\pi j\delta\mu$. This represents the number of particles (per unit area, time and energy) in a ring around $\boldsymbol{B}$, as shown in the sketch Fig. 4.6. One often leaves out the phase angle in j altogether, but this is dangerous because some 2π's may disappear in the process! One has to be very careful with the concept of gyrotropic fluxes. When there are drifts, a particle distribution will in general not be gyrotropic (just examine again Fig. 4.4b, c), even if the particle distribution is gyrotropic in the guiding center system.

If in addition to being gyrotropic the flux is independent of μ, $\partial j/\partial\mu = 0$, it is called *isotropic* (random pitch angles). For an isotropic flux, equal intervals of $\cos\alpha = \mu$ contain equal number of particles. This is the reason why pitch angle distributions are always plotted as a function of μ (not α)—an isotropic flux is represented by a horizontal straight line. Figure 4.7shows some typical pitch angle distributions of trapped particles, and their usual designations.

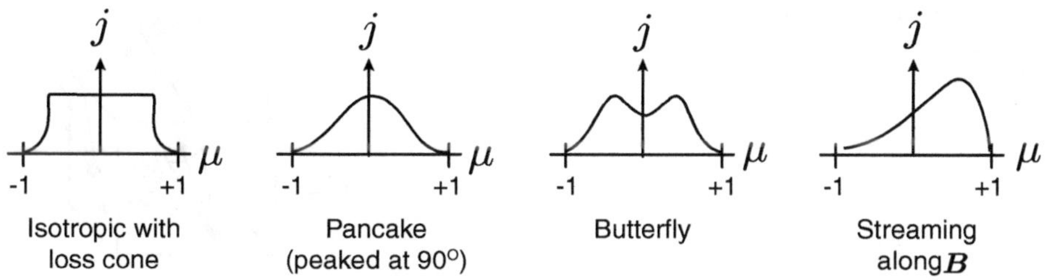

Fig. 4.7 Typical pitch angle distributions of trapped particles

4.2 Distribution Functions and Their Transformations

Measurements provide information about the flux $\boldsymbol{j}$. Theory, however, works with distribution functions, which serve as a quantitative bridge between the detailed but unknown instantaneous microscopic state of a particle ensemble and macroscopic variables more accessible to intuitive comprehension, theoretical treatment and measurement. Consider a small spatial volume δr^3 and call δn the number of particles in that volume whose momentum or velocity vectors fall into given ranges of magnitude and direction (black circles, Fig. 4.8) at a given time t.[2] In cartesian coordinates, that number will be proportional to δr^3 and to the element of volume δp^3 in momentum space, or δv^3 in velocity space, into which the pertinent vectors fall:

$$\delta n = f_p(\boldsymbol{r}, t, \boldsymbol{p})\delta r^3 \delta p_1 \delta p_2 \delta p_3 = f_v(\boldsymbol{r}, t, \boldsymbol{v})\delta r^3 \delta v_1 \delta v_2 \delta v_3 \tag{4.5}$$

f_v and f_p are *distribution functions*. Obviously, $f_p = f_v m^{-3}$. Note that f_p is the distribution function or *phase-space density* in the traditional six-dimensional phase space of statistical mechanics. The velocity space distribution function f_v can only be used in non-relativistic situations.

It often is convenient to work with other dynamical variables in the distribution function. Examples are (T, μ, φ); $(v_\perp, v_\parallel, \varphi)$; etc. Therefore, it is important to know the rule of transformation of the distribution function from a set of "old" variables, say v_1, v_2, v_3, to a "new" set X_1, X_2, X_3, each one of which is a function $X_i = X_i(v_1, v_2, v_3)$ of the three old variables (and, implicitly, space). The inverse transformations are $v_k = v_k(X_1, X_2, X_3)$. Obviously, we must have

$$\delta n/\delta r^3 = f_{new}(X_1, X_2, X_3)dX_1 dX_2 dX_3 = f_{old}(v_1, v_2, v_3)dv_1 dv_2 dv_3 \tag{4.6}$$

The integral over the dynamic variables X_i

$$n(\boldsymbol{r}, t) = \int_{X_i} f(\boldsymbol{r}, t, X_1, X_2, X_3)dX_1 dX_2 dX_3 \tag{4.7}$$

[2] All "delta" differentials in Fig. 4.8 (and in all that follows) are "physical" differentials as described in the footnote on page 185 in Appendix A.3: small, but still big enough to contain ⋙1 particles.

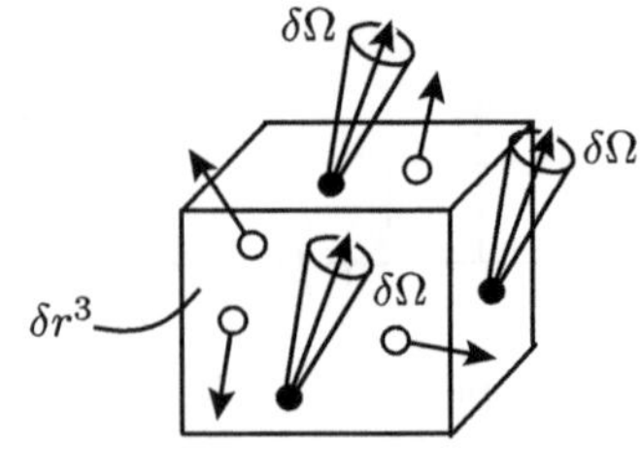

Fig. 4.8 Sketch to illustrate the definition of distribution function

is the *number density* of the particle ensemble at point $\boldsymbol{r}$ and time t (total number of particles per unit spatial volume at time t).

The corresponding elements of volume in (4.6) are related by the rule of the *Jacobian* $dX_1 dX_2 dX_3 = J_{XV} dv_1 dv_2 dv_3$ with the determinant (*not* a tensor!):

$$J_{XV} = J_{new/old} = \begin{vmatrix} \frac{\partial X_1}{\partial v_1} & \frac{\partial X_1}{\partial v_2} & \frac{\partial X_1}{\partial v_3} \\ \frac{\partial X_2}{\partial v_1} & \frac{\partial X_2}{\partial v_2} & \frac{\partial X_2}{\partial v_3} \\ \frac{\partial X_3}{\partial v_1} & \frac{\partial X_3}{\partial v_2} & \frac{\partial X_3}{\partial v_3} \end{vmatrix} = \left| \frac{DX}{DV} \right| \tag{4.8}$$

The inverse transformation $f_{new} \to f_{old}$ is given by the inverse J_{VX} of the Jacobian: $J_{VX} = J_{old/new} = |DV/DX| = 1/J_{XV}$, or $dv_1 dv_2 dv_3 = J_{VX}\, dX_1 dX_2 dX_3$. All this can be summarized by the relation

$$f_{new} = f_{old} \times J_{old/new} \tag{4.9}$$

in which the functions on the right side must be expressed *explicitly* in terms of the new variables, e.g.,

$$f_{old} = f_{old}[v_1(X_1, X_2, X_3), v_2(X_1, X_2, X_3), v_3(X_1, X_2, X_3)]$$

A confusing fact is that sometimes the relation (4.9) is written as $f_{new} = f_{old}$, attaching the Jacobian to the volume element whenever the latter appears explicitly. Also, in some books the Jacobian is defined with the transposed matrix of (4.8), with rows and columns exchanged. Note that when each one of the new variables is function of only one of the old variables ($X_k = X_k(v_k)$), the Jacobian is just the product $J_{new/old} = \partial X_1/\partial v_1 \cdot \partial X_2/\partial v_2 \cdot \partial X_3/\partial v_3$.

Let us discuss two important examples. First, the transformation from velocity space to the variables $[T, \mu, \varphi]$ in the natural frame of reference ($\boldsymbol{x} \parallel \boldsymbol{b}; \boldsymbol{y} \parallel \boldsymbol{n}; \boldsymbol{z} \parallel \boldsymbol{e}$ and $v_3 = v_\parallel$). We have

$$\begin{aligned} v_1 &= v\sqrt{1-\mu^2}\cos\varphi \\ v_2 &= v\sqrt{1-\mu^2}\sin\varphi \\ v_3 &= v\mu \end{aligned}$$

where $v = \sqrt{2T/m}$. The Jacobian $|D(v_1, v_2, v_3)/D(T, \mu, \varphi)|$ turns out to be $J_{old}/new = v/m = (\sqrt{2T/m})/m$; therefore

$$f_{new}(T, \mu, \varphi) = \frac{1}{m}\sqrt{\frac{2T}{m}} f_{old}(v_1, v_2, v_3)$$
$$= \frac{1}{m}\sqrt{\frac{2T}{m}} f_{old}\left(\sqrt{\frac{2T}{m}}\sqrt{1-\mu^2}\cos\varphi;\ \sqrt{\frac{2T}{m}}\sqrt{1-\mu^2}\sin\varphi;\ \sqrt{\frac{2T}{m}}\mu\right) \quad (4.10)$$

Observe how we have replaced in f_{old} the components of $\boldsymbol{v}$ by their functions of the new variables.

The other example is the transformation from a variable set $[v, \mu, \varphi]$ to the set $[v_\perp, v_\parallel, \varphi]$ we encountered in Sect. 2.3. The relationships are $v_\perp = v\sqrt{1-\mu^2}$, $v_\parallel = v\mu$ and $\varphi = \varphi$. The inverse Jacobian $J_{new/old}$ is easier to calculate:

$$J_{new/old} = \begin{vmatrix} \sqrt{1-\mu^2} & \mu & 0 \\ -\frac{v\mu}{\sqrt{1-\mu^2}} & v & 0 \\ 0 & 0 & 1 \end{vmatrix} = v/\sqrt{1-\mu^2} = 1/J_{old/new}$$

Calling F the new distribution function, we obtain, expressing v and μ in the new variables $v_\perp, v_\parallel$:

$$F(v_\perp, v_\parallel, \varphi) = \frac{v_\perp}{v_\perp^2 + v_\parallel^2} f\left(\sqrt{v_\perp^2 + v_\parallel^2};\ v_\parallel/\sqrt{v_\perp^2 + v_\parallel^2};\ \varphi\right) \quad (4.11)$$

The preceding discussion enables us to find the relation between particle flux $j(T, \mu, \varphi)$ and the corresponding distribution function. Consider a surface element $\delta A_\perp$ and δN particles crossing it perpendicularly during the interval δt with velocity $\boldsymbol{v} = v\boldsymbol{u}$ within an element $\delta\Omega$. The number ΔN of these particles is given by (4.1); prior to crossing, they all were contained in a cylinder perpendicular to the surface element of volume $\delta V = v\delta t\ \delta A_\perp$. Hence, by definition of distribution function we can also write $\Delta N = f(T, \mu, \phi)v\delta t\ \delta A_\perp \delta T, \delta\Omega$. Hence

$$j(T, \mu, \varphi) = v\ f(T, \mu, \varphi) \quad (4.12)$$

Based on (4.10) we then will have:

$$j(T, \mu, \varphi) = \frac{v^2}{m} f_v(v_1, v_2, v_3) = p^2 f_p(p_1, p_2, p_3) \quad (4.13)$$

This connects the phase space density (mostly wanted by theoreticians) with the particle flux (mostly provided by the experimentalists): $f_p = j/p^2$.

When a particle distribution is gyrotropic, the distribution functions do not depend on φ. But as stated in connection with Fig. 4.6, the integrations with

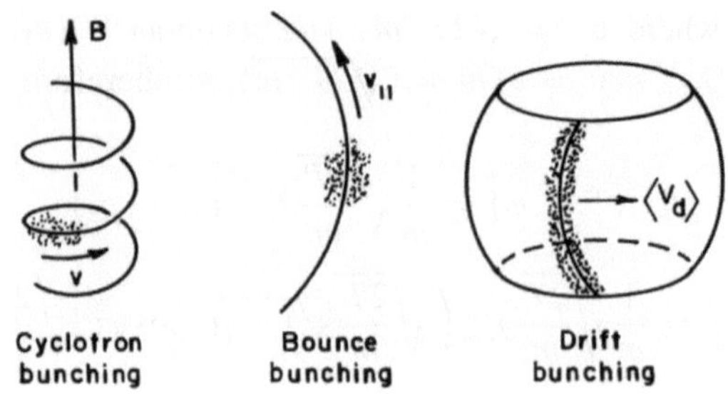

Fig. 4.9 Three cases in which the particle phase angles are not distributed equally (bunching)

respect to φ must not be forgotten, i.e., a factor 2π must always be added to the distribution function. Indeed, in practice trapped particle populations are gyrotropic, with equiprobable cyclotron phase angles that could be averaged out. Under stable magnetospheric conditions, during the lifetime of a trapped particle it will circle its drift shell many times so that, even if the particle population started as a compact *bunch* (e.g., in an artificial injection) (see Fig. 4.9) small initial differences in dynamic parameters as well as small time-dependent disturbances would smooth out the distribution throughout the entire shell.

4.3 Macroscopic Variables and the Particle Pressure Tensor

The distribution function is the "workhorse" of radiation belt and plasma theory. In the previous section we anticipated that it represents a bridge between the microscopic state of individual particles and macroscopic variables, which represent average effects of "zillions" of particles. The number density in (4.7) is an example. Quite generally, given a distribution function f_v, the average in velocity space of any microscopic variable $X(\boldsymbol{r}, t, \boldsymbol{v})$ will be

$$\langle X(\boldsymbol{r}, t)\rangle = \frac{\int X(\boldsymbol{r}, t, \boldsymbol{v}) f_v(\boldsymbol{r}, t, \boldsymbol{v}) dv_1 dv_2 dv_3}{\int f_v(\boldsymbol{r}, t, \boldsymbol{v}) dv_1 dv_2 dv_3} = \frac{1}{n}\int X f_v d^3v \tag{4.14}$$

In the following we list a series of such variables (called *velocity moments* when they involve velocity), which also play a fundamental role in plasma physics. f is the distribution function in velocity space $f_v(\boldsymbol{r}, t, v_1, v_2, v_3)$, but could be replaced by the distribution function in phase space f_p, provided that integrations are all carried out over $\boldsymbol{p}$.

Number density: $n(\boldsymbol{r}, t) = \int f\, d\boldsymbol{v}$

Mass density: $\rho_m(\boldsymbol{r}, t) = m\, n(\boldsymbol{r}, t) = m \int f\, d\boldsymbol{v}$

Average (or bulk) velocity: $\boldsymbol{V} = \langle \boldsymbol{v}\rangle = \int \boldsymbol{v} f\, d\boldsymbol{v} / \int f\, d\boldsymbol{v}$

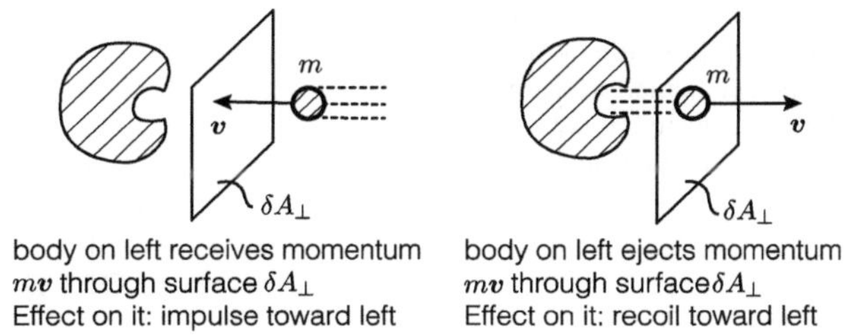

Fig. 4.10 Illustration showing why the effect of momentum transfer through a surface is independent of the direction of $\boldsymbol{v}$

Momentum density: $\boldsymbol{G} = nm\boldsymbol{V} = m \int \boldsymbol{v} f \, d\boldsymbol{v}$

Average kinetic energy density: $\epsilon = 1/2\, n\, m \langle v^2 \rangle = 1/2\, m \int v^2 f \, d\boldsymbol{v}$

Internal energy density (kinetic energy density in a frame traveling with bulk velocity $\boldsymbol{V}$): $w = 1/2\, m \int (\boldsymbol{v} - \langle \boldsymbol{v} \rangle)^2 f \, d\boldsymbol{v} = 1/2n\, m(\langle v^2 \rangle - \langle v \rangle^2) = \epsilon - 1/2\, n\, m\, V^2$

These relations are written for any frame of reference. In a local *natural* reference frame, the average or bulk velocity of an ensemble of adiabatically moving charged particles is composed of the ensemble-average drift velocity perpendicular to the magnetic field and the ensemble-average parallel velocity along it. Under adiabatic conditions, these are also the average guiding center velocity components.

To the above list we must add an important macroscopic quantity, the *pressure tensor*. First, let us remember that the primary, kinetic, concept of pressure is related to momentum transfer across a surface. Consider Fig. 4.10: a body on the left side either receives or ejects a little mass m that traverses a perpendicular surface $\delta A_\perp$. The picture shows that the effect on the body will be the same, regardless of the sense of motion of the little mass through the surface. Trivially (but sometimes forgotten): a gain of momentum *from* the right has the same effect as a loss of momentum *to* the right. If there are many particles (with number density n and the same mass and velocity perpendicular to $\delta A_\perp$) the flux of momentum per unit area and time will be $(nmv) \times v$; again, independent of the sense of $\boldsymbol{v}$.

Second, a "kindergarten" example will show us *why* at all there is pressure in an adiabatic ensemble of charged particles, even in the case when they all have the same kinetic energy and pitch angle. We consider the ubiquitous example of a gyrotropic distribution of 90° pitch angle particles in a uniform magnetic field, in absence of other forces (Fig. 4.11). If we cut this distribution with a surface (perpendicular to the paper) each circling particle whose guiding center lies within $2\rho_C$ of the x-line will transfer a perpendicular momentum $mv_\perp \sin\theta$ every turn (of duration $\tau_C = 2\pi m/qB$—note that there are two angles θ per turn!). There are $ndy\Delta x\Delta z$ guiding centers in the hatched band $dy = \rho_C \sin\theta d\theta$. Therefore, the total perpendicular momentum flux $\Delta G_\perp$ per unit area and time will be

$$\frac{\Delta G}{\Delta x \Delta z} = p_\perp = \frac{nmv_\perp^2}{2\pi} \int_0^{2\pi} \sin^2\theta d\theta = 1/2nmv_\perp^2$$

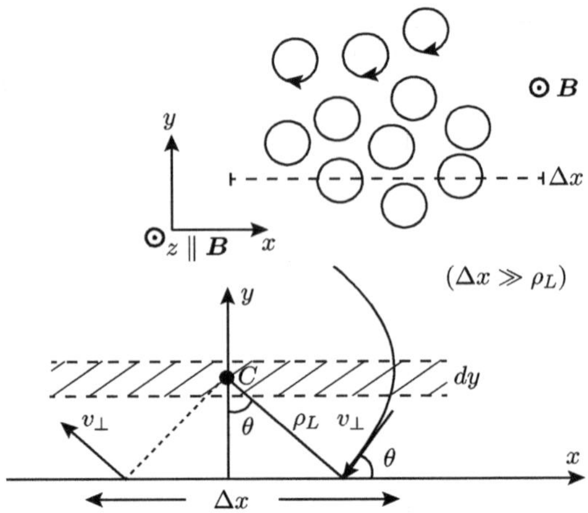

Fig. 4.11 Cyclotron orbit geometry illustrating how the perpendicular pressure on the $x - z$ plane arises from the cyclotron motion

Clearly, this perpendicular pressure (see below) arises from the fact that the non-drifting ensemble "rattles" on any surface perpendicular to their cyclotron motion. Even if all particles have the same velocity, as happens in this example, it is the collective macroscopic action of their random cyclotron phases that is represented in the pressure variable. A similar consideration can be made if particles also have a parallel velocity $v_\parallel$ ($\alpha \neq 90°$); in that case the momentum flux parallel to $\boldsymbol{B}$ would be $\Delta G_\parallel/\Delta x \Delta y = p_\parallel = nmv_\parallel^2$.

Finally, the perpendicular way is not the only way to transport momentum through a surface: *horizontal* momentum mv_x can be transported vertically through a horizontal surface δA (think of what happens when you are cleaning the deck with a water hose: you are transferring vertically downwards the horizontal momentum needed to push the dirt out of the way!). For a large ensemble a downward horizontal momentum transfer in the x-direction would be $(nmv_x) \times v_z$. Momentum transfer quantities are thus components of a tensor (Appendix A.1). If we have a whole velocity distribution of particles, we define the *kinetic tensor* as $\mathbb{K} = n\langle \boldsymbol{v} \otimes \boldsymbol{p}\rangle = m \int \boldsymbol{v} \otimes \boldsymbol{v} f_v(\boldsymbol{v}) d^3v$, with components

$$\mathbb{K}_{ik} = \int v_i p_k f_p(\boldsymbol{p}) d^3p = m \int v_i v_k f_v(\boldsymbol{v}) d^3v \tag{4.15}$$

Note that $\mathbb{K}$ is a symmetric tensor, so that only six components are independent. The product $\boldsymbol{\delta} G = \mathbb{K}\boldsymbol{\delta} A$ ($\delta G_i = \sum_k \mathbb{K}_{ik} \delta A_k$) is the momentum flow through the element of area $\boldsymbol{\delta} A$.

More attune to our concept of pressure would be a physical entity whose components represent momentum transfer in a frame of reference that locally moves with the bulk velocity $\boldsymbol{V}$ of the particle ensemble. Like in thermodynamics this would give us a better picture of the "randomness" of the motion of individual particles and potential effects on macroscopic bodies or obstacles like walls. We thus define the *pressure tensor* as

$$\mathbb{P} = m \int (v - V) \otimes (v - V) f_v(\boldsymbol{v}) d^3 v = \mathbb{K} - \mathbb{D} \tag{4.16}$$

where $\mathbb{D} = nm\boldsymbol{V} \otimes \boldsymbol{V}$ is the *dynamic pressure tensor*, representing the contribution to the momentum transfer of the common translational motion of the ensemble.

Let us now examine an example in the natural frame of reference. We shall consider a gyrotropic ensemble in a uniform magnetic field and no forces, with a distribution function $F(v_\perp, v_\parallel)$ like in (4.11), but independent of φ (a fact which, nevertheless, must not be forgotten in the integrals in phase space!) If the x_3 axis is $\parallel \boldsymbol{B}$ (unit vector $\boldsymbol{e}$) and the other axes x_1, x_2 are $\boldsymbol{b}$ and $\boldsymbol{n}$ respectively (see Fig. A.1 of Appendix A.1), the vector components of $\boldsymbol{v}_\perp$ will be $v_\perp \cos\varphi$ and $v_\perp \sin\varphi$, respectively. Therefore, in the expression (4.15) of the kinetic tensor components, all integrals over the cyclotron phase angle that have $\sin\varphi, \cos\varphi, \sin\varphi\cos\varphi$ will be zero. This leaves only the diagonal components (with the squares of sin and cos), which turn out to be $\mathbb{K}_{11} = \mathbb{K}_{22} = 1/2nm\langle v_\perp^2\rangle$ and $\mathbb{K}_{33} = nm\langle v_\parallel^2\rangle$. In these expressions, the density is $n = \int F dv_\perp dv_\parallel d\varphi = 2\pi \int F dv_\perp dv_\parallel$. Taking into account (4.16), we finally have for the pressure tensor in the natural coordinate system:

$$\begin{aligned}
\mathbb{P} &= \begin{pmatrix} p_\perp & 0 & 0 \\ 0 & p_\perp & 0 \\ 0 & 0 & p_\parallel \end{pmatrix} \\
p_\perp &= \frac{1}{2} nm\left[\langle v_\perp^2\rangle - \langle \boldsymbol{v}_\perp\rangle^2\right] = \frac{1}{2} nm\left[\langle v_\perp^2\rangle - V_D^2\right] \\
p_\parallel &= nm\left[\langle v_\parallel^2\rangle - \langle \boldsymbol{v}_\parallel\rangle^2\right] = nm\left[\langle v_\parallel^2\rangle - V_\parallel^2\right]
\end{aligned} \tag{4.17}$$

Notice that the scalar *trace* (Appendix A.1) of the pressure tensor is related to the internal energy density:

$$w = \frac{1}{2} \mathrm{Tr}\, \mathbb{P}$$

When the pitch angle distribution is independent of the particle's energy, i.e., when we have separation of variables in the distribution function $f = h(T)g(\mu)$, we'll have $\langle v_\perp^2\rangle = \langle v^2\rangle(1 - \langle \mu^2\rangle)$, $\langle v_\parallel^2\rangle = \langle v^2\rangle\langle \mu^2\rangle$ and $\langle v_\parallel\rangle^2 = \langle v\rangle^2\langle \mu\rangle^2$ (where the averages are taken over the respective variables). Neglecting the average drift velocity V_D, relations (4.17) become

$$\begin{aligned}
p_\perp &= \frac{1}{2} nm\langle v^2\rangle\left[1 - \langle \mu^2\rangle\right] \\
p_\parallel &= nm\langle v^2\rangle\left[\langle \mu^2\rangle - \langle \mu\rangle^2\right]
\end{aligned} \tag{4.18}$$

For an isotropic distribution, $\langle \mu\rangle = 0$ and $\langle \mu^2\rangle = \int_{-1}^{+1} \mu^2 d\mu / \int_{-1}^{+1} d\mu = 1/3$, so that

$$p_\perp = p_\parallel = p = \frac{1}{3}nm\langle v^2\rangle \tag{4.19}$$

is the *isotropic* pressure, a scalar that is the common diagonal element of the tensor $\mathbb{P}$, which in this case can be written as $\mathbb{P} = p\,\mathbb{I}$, with $\mathbb{I}$ the unit tensor of elements $\delta_{ik} = 1$ if $i = k$ and zero otherwise.

Taking into account that the tensor square of the unitary vector $\boldsymbol{e} = \boldsymbol{B}/B$ in the natural coordinate system is represented by a matrix with zeros everywhere except 1 at $i, j = 3, 3$, we can write the following for the pressure tensor (4.17):

$$\mathbb{P} = p_\perp\mathbb{I} + (p_\parallel - p_\perp)\boldsymbol{e} \otimes \boldsymbol{e} = p_\perp\mathbb{I} + (p_\parallel - p_\perp)(\boldsymbol{B} \otimes \boldsymbol{B}/B^2) \tag{4.20}$$

The last expression is useful when the components of $\mathbb{P}$ have to be transformed to other coordinate systems. In particular, a quantity that is important in plasma physics is the *divergence* of tensor $\mathbb{P}$, of components (in any coordinate system) $\nabla\mathbb{P}|_i = \Sigma_k \partial P_{ik}/\partial x_k$ (see Appendix A.1). Manipulating the second part of Eq. (4.20) by components, one obtains (rather laboriously) the following expressions for the perpendicular and parallel components of the divergence:

$$\begin{aligned} \nabla\mathbb{P}|_\perp &= \nabla_\perp p_\perp + \left(p_\parallel - p_\perp\right)\frac{\partial \boldsymbol{e}}{\partial s} \\ \nabla\mathbb{P}|_\parallel &= \frac{\partial p_\parallel}{\partial s} + \left(p_\perp - p_\parallel\right)\frac{1}{B}\frac{\partial B}{\partial s} \end{aligned} \tag{4.21}$$

Some may wonder: how come a purely "mechanical" quantity like the divergence of the pressure tensor depends on the magnetic field? Certainly, the pressure tensor *per se* as defined in (4.16) is independent of $\boldsymbol{B}$, but the above relations (and even its form (4.17)) are representations in the natural coordinate system which *does* depend on the magnetic field geometry (see Fig. A.1) and which for an inhomogeneous field varies from point to point. In this context, notice the following important facts: (i) if the particle distribution is fully isotropic, $p_\parallel = p_\perp$, the terms containing the magnetic field in Eqs. (4.21) are indeed zero and $\nabla\mathbb{P} = \nabla p$; (ii) in the general case, only the field *geometry* intervenes: multiplying the magnetic field magnitude by a number everywhere does not change the values of the last terms in (4.21).

With the pressure tensor $\mathbb{P}$ one can define a *temperature tensor* $\mathbb{T}$ in the form

$$\mathbb{T} = \frac{1}{nk}\mathbb{P}$$

where $k = 1.38 \times 10^{-23}$ Joule/°K is Boltzmann's constant. In case of isotropy, we have a scalar temperature T (do not confuse with kinetic energy!) $nkT = p = 1/3nm\langle v^2\rangle$, or

$$kT = 1/3m\langle v^2\rangle = m{v_{th}}^2 \tag{4.22}$$

where v_{th} is called the *thermal speed*. For a *gyrotropic* distribution (not necessarily isotropic) we have two types of temperatures:

$$\begin{aligned} T_\perp &= \frac{1}{nk} p_\perp = \frac{1}{2k} m\left[\langle {v_\perp}^2 \rangle - \langle v_\perp \rangle^2\right] \\ T_\parallel &= \frac{1}{nk} p_\parallel = \frac{1}{k} m\left[\langle {v_\parallel}^2 \rangle - \langle v_\parallel \rangle^2\right] \end{aligned} \tag{4.23}$$

Note that in the case of a *monoenergetic, mono-pitch-angle* distribution of particles, the parallel temperature would be zero but we *still* would have a perpendicular temperature $T_\perp \neq 0$! This perpendicular temperature arises from the randomness of the particles' cyclotron motion (page 98). As we decrease gyrotropicity (bunching of the cyclotron motion, Fig. 4.9 left), the perpendicular temperature will decrease until it hits zero when *all* particles gyrate in phase.

The final point of this section concerns the magnetization density of an ensemble of adiabatic charged particles. In the guiding center particle model, each particle has an intrinsic magnetic moment given by (1.26) in which, as we must remember, the transverse velocity or kinetic energy is that of the particle in the GCS: $\boldsymbol{M} = -m(\boldsymbol{v}_\perp - \boldsymbol{V}_D)^2/(2B)\,\boldsymbol{e}$. Averaging over all particles in an element of volume, and taking into account the definition (4.17), we obtain the following expression for the magnetic moment *density* (which henceforth we will designate with M, hoping that there will not be too much confusion with the magnetic moment of an individual particle):

$$\boldsymbol{M} = -\frac{p_\perp}{B}\boldsymbol{e} \tag{4.24}$$

4.4 Liouville's Theorem and Stationary Trapped Particle Ensembles

There is a general theorem which can be derived directly from Hamiltonian theory, valid for any ensemble of collisionless particles subjected to non-dissipative forces: *Liouville's Theorem* states that the distribution function along the trajectory through any point $\boldsymbol{r}, \boldsymbol{p}$ in phase space[3] is constant. In other words, nearby particles in phase space "stick together" in such a way that as time evolves, their local phase space density f_p (*not* to confuse with the ordinary number density $n(\boldsymbol{r}, t) = \int f_p \, d\boldsymbol{p}$) does not change. This means that for the total time derivative:

$$\frac{df_p}{dt} = \frac{\partial f_p}{\partial t} + \frac{d\boldsymbol{r}}{dt}\cdot\nabla_r f_p + \frac{d\boldsymbol{p}}{dt}\cdot\nabla_{\boldsymbol{p}} f_p = 0 \tag{4.25}$$

[3]Remember that through each point in phase space there is only *one* trajectory for a given force field.

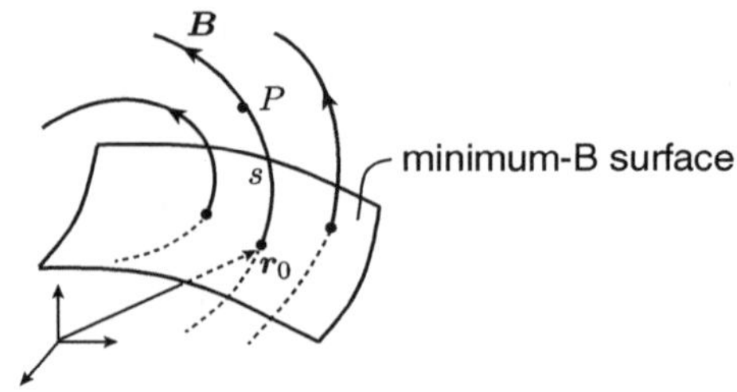

Fig. 4.12 Space variables used in the example

In this expression (non-relativistic case!) $d\boldsymbol{r}/dt = \boldsymbol{p}/m$, and $d\boldsymbol{p}/dt = \boldsymbol{F}$ is the common acceleration caused by the local force field $\boldsymbol{F}$ acting on the particles in the infinitesimal neighborhood of $(\boldsymbol{r}, \boldsymbol{p})$. The vector operator ∇_r has components $\partial/\partial x_i$, those of ∇_p are $\partial/\partial p_i$. It is useful to point out that for a magnetic flux tube filled with streaming particles, it is possible to demonstrate the constancy of j/p^2 (see (4.13)) *without* invoking Hamilton's equations.

To achieve a more practical understanding of some of the relations presented in the preceding section, let us examine several concrete examples. First we consider an ensemble of identical particles trapped on field lines under adiabatic conditions—for instance, radiation belt protons. We still uphold some basic restrictions like absence of field-aligned electric fields, no time-variations, and symmetric angular distributions everywhere so that no bounce-phase bunching occurs. Finally, we make another fundamental, though not too unrealistic, assumption that the distribution function of trapped particles is separable in the form:

$$F(\boldsymbol{r}, T, \mu) = h(\boldsymbol{r}, T)g(\boldsymbol{r}, \mu) = h(\boldsymbol{r}_0, T)g(\boldsymbol{r}_0, s, \mu)$$

In the second equality, we have chosen three coordinates related to the so-called α, β system (see Fig. 3.2): the position of the intersection point of the field line with the minimum-B surface (denoted by the vector $\boldsymbol{r}_0(\alpha, \beta)$), plus the curvilinear variable s along the field line (see Fig. 4.12). Since the field lines were assumed to be electric equipotentials, all adiabatic relations linking field-aligned quantities are energy-independent and the functional form of the energy spectrum is the same for all particles s along a field line. Moreover, since initially we will only be interested in the dynamic relationships between particle density and pitch angle distribution along just *one* given field line, we choose the notation

$$F(s, T, \mu) = h(T)\, g_s(\mu_s) \tag{4.26}$$

Again, $h(T)$ is the common energy spectral form on the field line and $g_s(\mu_s)$ the pitch angle distribution function at point s of the field line. Note carefully that because of our a priori restrictions, all pitch angle distributions must be symmetric: $g_s(\mu_s) = g_s(-\mu_s)$. With this notation, the integral particle density at the equatorial point will be $n_0 = \int_0^\infty h(T)dT$ and the average (non-relativistic) particle kinetic energy $\langle T \rangle = \int_0^\infty T h(T)dT$ (a quantity that is constant along the field line).

It is clear that the pitch angle distribution $g_s(\mu_s)$ must be related to that at the minimum-B point, $g_0(\mu_0)$, because all particles that traverse point s must pass

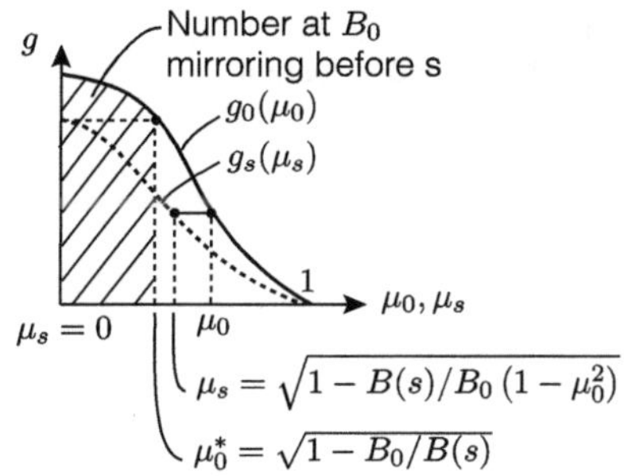

Fig. 4.13 Sketch of a pitch angle dependence of distribution function at the equatorial point (g_0) and on a generic field line point (g_s)

through $\boldsymbol{r}_0$. The relation comes from Liouville's equation (4.25) (see also Fig. 2.6), and is simply written in terms of the pitch angle distribution functions:

$$g_s(\mu_s) = g_0(\mu_0)$$

The pitch angle μ_s is the pitch angle at point s of a particle that has a pitch angle μ_0 at the equator. The two pitch angles are related by (2.38):

$$\mu_s = \sqrt{1 - \frac{B(s)}{B_0}(1 - {\mu_0}^2)} \quad \text{or} \quad \mu_0 = \sqrt{1 - \frac{B_0}{B(s)}(1 - {\mu_s}^2)} \tag{4.27}$$

Given the pitch angle distribution $g_0(\mu_0)$ at the equatorial point, at any other off-equatorial point of the field line

$$g_s(\mu_s) = g_0\left(\sqrt{1 - \frac{B_0}{B(s)}(1 - {\mu_s}^2)}\right) \tag{4.28}$$

Figure 4.13 is useful to interpret the relationship, and it allows to construct graphically the pitch angle distribution at any point of a field line, given the one at the minimum-B point. Particles with μ below the limiting value $\mu_0^* = \sqrt{1 - B_0/B(s)}$ at the equator mirror before reaching point s. Concerning the loss cone (2.42), we assume it to be included in the functional form of g_0.

Next we examine the total particle density along a field line. At point s it will be:

$$n(s) = \int_0^\infty h(T)dT \int_{-1}^{+1} g_s(\mu_s)d\mu_s = n_0 \int_{-1}^{+1} g_0(\mu_0)\frac{d\mu_s}{d\mu_0}d\mu_0$$

In this relation, $d\mu_s/d\mu_0$ is obtained from (4.27). As an integral over μ_0, we can go only from -1 to $-\mu_0^*$, and then from $+\mu_0^*$ to +1 (for the definition of μ_0^*, observe carefully Fig. 4.13!); both sections are equal because of the symmetry of the function $g(\mu)$. This leaves us with

$$n(s) = n_0\,\frac{B(s)}{B_0}\,2\int_{\mu_0^*}^1 g_0(\mu_0)\frac{\mu_0 d\mu_0}{\sqrt{1 - (B(s)/B_0)(1 - \mu_0^2)}} \tag{4.29}$$

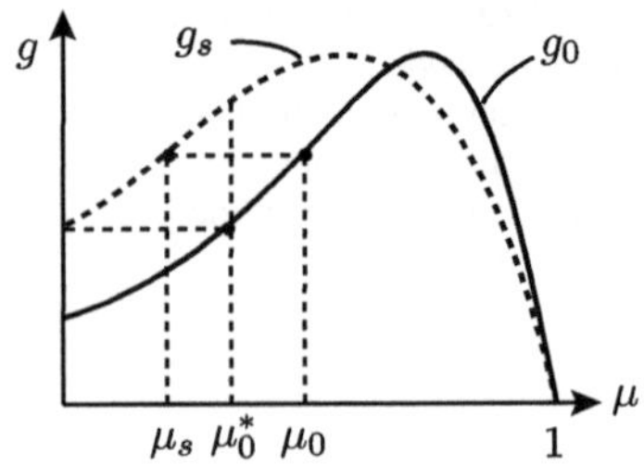

Fig. 4.14 Same as Fig. 4.13, for a butterfly distribution

Given the pitch angle distribution $g_0(\mu_0)$ and integral density n_0 at the equatorial point of an equipotential field line, this equation gives us the density anywhere along that field line. And conversely, given the density function of a trapped particle population as a function of s along a magnetic field line, it is possible to find the pitch angle distribution at the equatorial point $g_0(\mu_0)$ as the inverse solution of (4.29) (a *Volterra-Abel* integral equation). This may be useful for experimentalists when the only available source of data are integral density or flux measurements with omnidirectional detectors.

We can extract further important facts from Eq. (4.29). First, for an *isotropic* distribution ($g_0 = \text{const.}$) the integral density $n(s)$ will be constant along the field line (verify analytically, or graphically using Fig. 4.13). This seems contradictory: as we move away from the minimum-B point, should we not be losing the particles which mirror on the way? We will, indeed, but as B increases, the flux tube narrows down and the remaining particles get compressed in exactly such a way that for an isotropic angular distribution the particle *density* remains constant! Second, for a *pancake* distribution (peak at $\alpha = 90°$) a graphic examination of Fig. 4.13 shows that the number density will decrease as we move away from the equator. Third, in contrast, a similar graphic examination shows that for a *butterfly* distribution (dip at 90°, Fig. 4.14) the integral particle density will *increase* as we move away from the equator.

Next, we examine the particle pressures along a field line as a function of the pitch angle distribution at the equatorial point. Starting with the definitions (4.15) and (4.16), and taking into account that in our case the ensemble-average (bulk) velocities $\langle V_\perp \rangle$ and $\langle V_\parallel \rangle$ are both zero, we have, at point s:

$$p_\perp(s) = \frac{1}{2}\int_0^\infty mv^2 h(T)dT \int_{-1}^{+1} (1-\mu_s^2) g_s(\mu_s) d\mu_s$$

Taking into account the integral transformations leading to (4.29) and realizing that $\int_0^\infty 1/2mv^2 h(T)dT = \langle T \rangle$, the constant average particle energy, we end up with

$$p_\perp(s) = 2\langle T \rangle \left(\frac{B(s)}{B_0}\right)^2 \int_{\mu_0^*}^1 (1-\mu_0^2) g_0(\mu_0) \frac{\mu_0 d\mu_0}{\sqrt{1-(B(s)/B_0)(1-\mu_0^2)}} \tag{4.30}$$

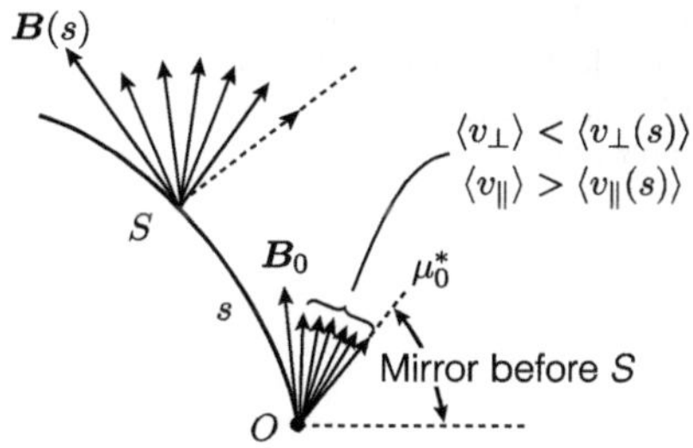

Fig. 4.15 Relation between pitch angles on the equator and a generic point s

Likewise, for the parallel pressure we obtain:

$$p_\parallel(s) = 4\langle T\rangle \frac{B(s)}{B_0} \int_{\mu_0^*}^{1} g_0(\mu_0)\sqrt{1-(B(s)/B_0)(1-\mu_0^2)}\mu_0 d\mu_0 \qquad (4.31)$$

At any point s the two pressures are mathematically related. If we laboriously but straightforwardly evaluate the partial derivative $\partial/\partial s(p_\parallel/B(s))$ using Eq. (4.31) and take into account (4.30), we obtain

$$B(s)\frac{\partial}{\partial s}\left(\frac{p_\parallel}{B(s)}\right) + \frac{p_\perp}{B(s)}\frac{\partial B}{\partial s} = \frac{\partial p_\parallel}{\partial s} + \left(p_\perp - p_\parallel\right)\frac{1}{B}\frac{\partial B}{\partial s} = 0 \qquad (4.32)$$

A similar procedure for $\nabla_\perp p_\perp$ leads to relation

$$\nabla_\perp p_\perp + (p_\parallel - p_\perp)\frac{\partial \boldsymbol{e}}{\partial s} = 0 \qquad (4.33)$$

These indeed correspond to the expressions of the parallel and perpendicular components of the divergence of tensor $\mathbb{P}$ (Eqs. (4.21)) when there is equilibrium: $\nabla\mathbb{P} = 0$.

A discussion of some special questions is in order. First, consider an *isotropic* distribution at the equatorial point ($g_0(\mu_0)$ =const.). In that case, relations (4.28), (4.30), (4.31) and Fig. 4.13 show us that the pitch angle distribution will be isotropic at all points of the field line, and $p_\perp = p_\parallel = p$ =const. An apparent paradox arises: If p is independent of s, since B increases as we move away from the minimum-B point, this means that the average magnetic moment density $M = p_\perp/B$ will decrease. The magnetic moment of individual particles *is* conserved—Shouldn't one expect the ensemble-average M also to be conserved? Examine Fig. 4.15 and consider all particles at point S (with pitch angles between 0 and 90°). When they pass through the equatorial point O they form a sub-population contained in a solid angle cone shown in the figure. Clearly, their average *perpendicular* velocity $\langle v_{\perp 0}^2\rangle$ at that point is less than $\langle v_{\perp s}^2\rangle$, in order to conserve their individual magnetic moments $T_\perp/B$. It is the *additional* particles at the equatorial point O, filling the rest of the pitch angle range and which mirror *before* they can reach point s, which contribute to an *increase* of the ensemble-average $\langle v_\perp\rangle$ at that point, so as to keep it and the perpendicular pressure constant at all points (and cause a change in

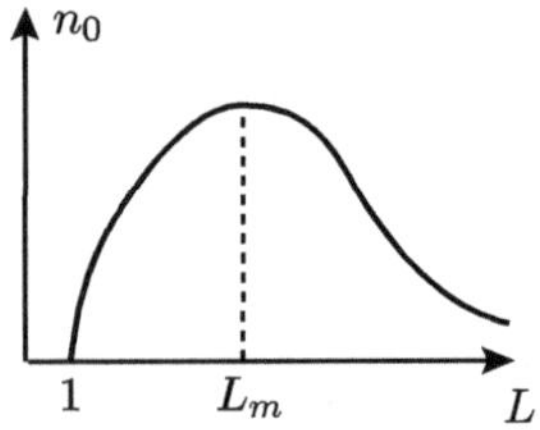

Fig. 4.16 Sketch of the particle density as a function of L at the equator in a dipole field

the magnetic moment *density* M along the field line). No adiabatic conservation theorem is violated! To make this picture even a little more complicated, let us mention another apparent paradox. If because of the individual conservation of magnetic moment the sub-population of Fig. 4.15 increases its average $\langle v_\perp \rangle$ as it moves away from O, its average $v_\parallel$ must decrease because of the conservation of kinetic energy (see the projections of the vectors $\parallel$ to $\boldsymbol{B}$). How come, then, does not the parallel pressure decrease, either? The explanation is a bit more complicated, but is based on the same argument as before: the parallel pressure of the *total* particle population is what remains constant along a field line! Remember that all this was shown for the special case of an isotropic distribution of trapped particles, separable in energy, and an equilibrium situation.

For the last example we will restrict the magnetic field to that of a pure dipole field, a reasonable approximation for inner radiation belt. But now the discussion will not be limited to just one field line—we will first examine what happens on one meridian, and then extend the picture into three dimensions (albeit cylindrically symmetric). Again, refer to relations (3.22)–(3.28) for the most important geometric features and relationships. Figure 3.14 reminds the reader of the symbols used. We are interested in finding the relationships between variables like integral particle density n, and the pressures $p_\perp$ and $p_\parallel$ at any point B, L of the dipole field, given the (thus far arbitrary) particle density at the equatorial surface $n_0(L)$. For our semi-quantitative analysis, we shall adopt as integral density $n_0(L)$ a function of the type shown in Fig. 4.16, and assume, as above, a separable distribution function like (4.26). As to the pitch angle distribution at the minimum-B surface we shall assume that it is isotropic everywhere, but with a local loss cone cosine μ_L given by (3.28): $g_0(\mu_0) = g_0$ if $|\mu_0| \le \mu_L$ and $g_0(\mu_0) = 0$ for $\mu_L < |\mu_0| \le 1$. The particle density at any point L, s of the dipole field will be given by (4.29), which in this particular case turns out to be

$$n(L,s) = \int_0^\infty h(T)dT \int_{-\mu_L(s)}^{+\mu_L(s)} g_0 d\mu_s = n_0(L)\frac{\mu_L(s)}{\mu_L}$$

If instead of s we choose the local magnetic field B, we obtain quite simple relations:

$$n(L,B) = n_0(L)\sqrt{\frac{1 - B/B_L}{1 - B_0/B_L}} \tag{4.34}$$

Figure 4.17 shows qualitatively the density dependence along a dipole field line between the minimum-B point and the Earth intersection for an isotropic pitch

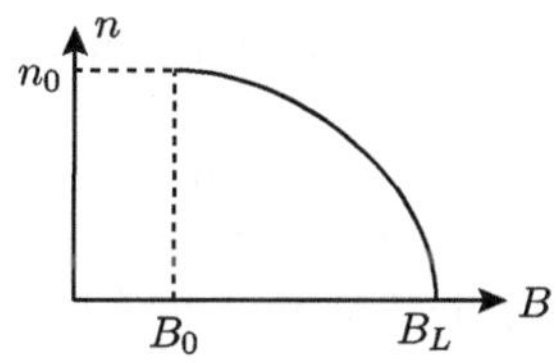

Fig. 4.17 Sketch of particle density along a field line for an isotropic distribution at the equator cut off by the loss cone

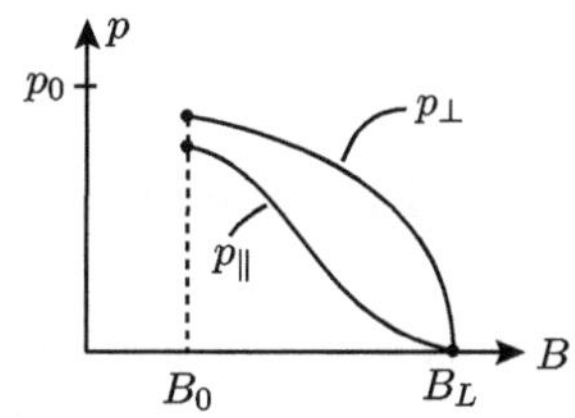

Fig. 4.18 Sketch of the corresponding parallel and perpendicular pressures

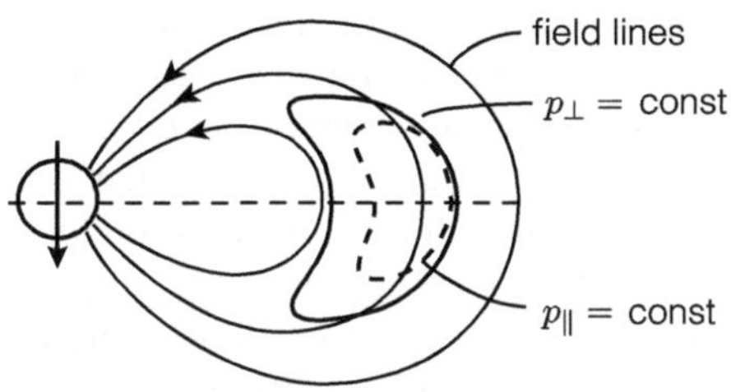

Fig. 4.19 Sketch of parallel and perpendicular isobars on a meridional plane

angle distribution (with loss cones) at the equator. The variation of this otherwise isotropic angular distribution is due entirely to the ever-increasing loss cone as one approaches the Earth. For the pressures we only need their original definition (4.17), and figure out the averages for an isotropic pitch angle distribution cut off at the loss cone cosine μ_L. We obtain

$$p_\perp(L, B) = \frac{2}{3} n_0(L)\langle T\rangle \sqrt{\frac{1 - B/B_L}{1 - B_0/B_L}} \left(1 - \frac{1}{2}\frac{B}{B_L}\right) \tag{4.35}$$

and

$$p_\parallel(L, B) = \frac{2}{3} n_0(L)\langle T\rangle \sqrt{\frac{1 - B/B_L}{1 - B_0/B_L}} \left(1 - \frac{B}{B_L}\right) \tag{4.36}$$

Note that $2/3\, n_0(L)\langle T\rangle$ is the isotropic pressure p that would exist on the equator if the loss cone were exactly = 0 ($\mu_L = 1$). As a function of B along a dipole field line, the pressures vary as sketched in Fig. 4.18. It is important to observe how each type of pressure approaches zero (their second derivatives) at the Earth intersection (B_L). Contours of constant $p_\perp$ and $p_\parallel$ in B, L space, a frequent representation of radiation belt intensity, are given by (4.35) and (4.36) and sketched in Fig. 4.19.

Observe carefully how they are tangent to the field lines close to the equatorial points.

It is important to remember that in the preceding discussion we were dealing with a stationary situation. In that case, the entire trapped particle population is governed by the function $n_0(L)$—but we don't really know yet *how this function is determined* for an equilibrium situation! Dynamic processes take care of that (Chap. 5).

4.5 Particle Distributions and Mapping in Invariant Space

Let us return to the general distribution function in (4.7), taking as its variables (X_1, X_2, X_3) the three adiabatic invariants of a particle pertaining to an ensemble trapped in a given magnetic field $\boldsymbol{B}(\boldsymbol{r}, t)$ (and eventually, electric field $\boldsymbol{E}(\boldsymbol{r}, t)$). It is advisable to have a clear understanding of the physical meaning of the *theoretically important invariant space* M, J, Φ, the distribution of guiding center particles in invariant space (which we now will designate with $F(M, J, \Phi)$), the comparison with other distribution functions, and the relation to the *experimentally accessible particle flux* $j(\boldsymbol{r}, T, \mu, \varphi)$ (or $2\pi j(\boldsymbol{r}, T, \mu)$ if gyrotropic). A point in invariant space represents one entire drift shell, limited by the mirror point traces (see shell Φ in Fig. 4.20), engendered by a specific class of particles, *regardless* of their actual cyclotron, bounce and drift phases in their adiabatic motion (we are assuming no bunching in any of the periodic motions). The quantity $\delta N = F(M, J, \Phi)\delta M\, \delta J\, \delta\Phi$ represents the total number of guiding center particles with magnetic moment lying between M and $M + \delta M$ and second invariant value between J and $J + \delta J$, which are found anywhere in longitude and latitude on shells (note plural!) whose third invariant value lies between Φ and $\Phi + \delta\Phi$, as shown in Fig. 4.20. $F(M, J, \Phi)$ is, therefore, a guiding center particle *number density* in invariant space. Again, there is no mention of particular phases of real particle periodic motions. Similar to phase space in statistical mechanics, one single point in the adiabatic invariant space represents one specific but very complex configuration of particles (their guiding centers) in real space.

Let us discuss an extreme example: when the invariant space distribution function is a delta function: $F(M, J, \Phi) = N\delta[(M - M_0), (J - J_0), (\Phi - \Phi_0)]$. This means that we have an ensemble of N particles, all with the *same* magnetic moment and J-value with guiding centers on a *single* shell of flux Φ_0. The corresponding density function in invariant space is an integrable singularity on a single point (M_0, J_0, Φ_0) of invariant space. In real space, it means that an ideal directional differential detector (page 89) will only register particles if placed on the right drift shell (Fig. 4.21), looking into the right direction with the right energy spectrum window (both extractable from the local expressions of M and J). Now assume that the fields change slowly, adiabatically (relations (2.1) and (2.2)), so that all three invariants are conserved, and consider (A.74) and the last paragraphs of Appendix A.3. *On the average*, the place (M_0, J_0, Φ_0) of the delta function will remain the same—but looking with a magnifying glass into invariant space we will

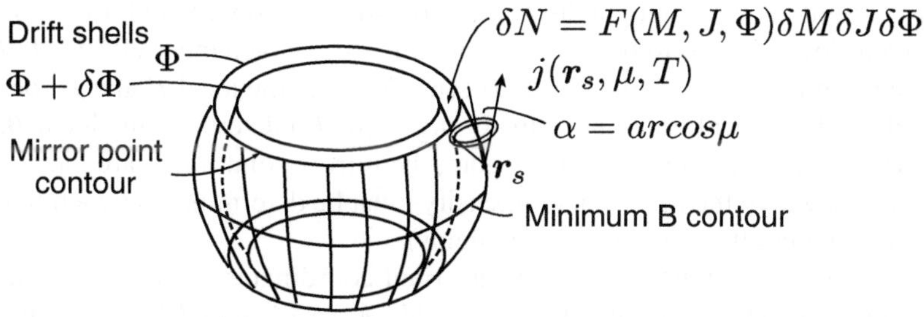

Fig. 4.20 Sketch of drift shells Φ and $\Phi + \delta\Phi$ (in a dipole-like trapping field the third invariant is defined by the magnetic flux *outside* the shell; Fig. 3.22). δN is the number of guiding center particles with the given M, J and Φ values inside the space between shells, at a given time. The little cone shown on the right is really the same as in Fig. 4.6: $j\delta A_{\perp}\,\delta\mu\,\delta T$ is the number per unit time of the "real" particles in that space which cross the surface element perpendicular to their motion at shell point $\boldsymbol{r}$ and the μ and T values *there*

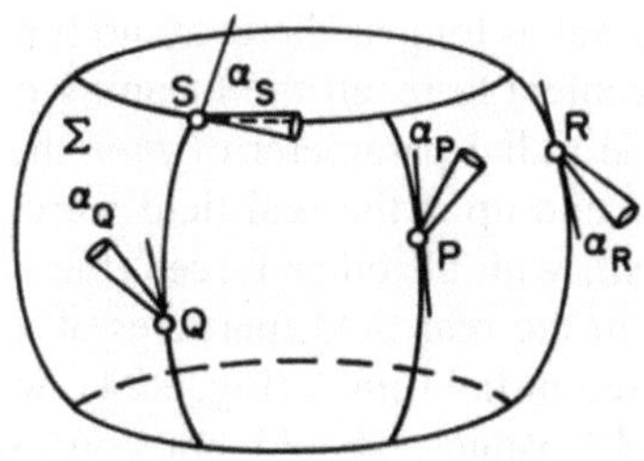

Fig. 4.21 Drift shell of particles of the same values of M_0, J_0 and Φ_0 and sketch of their flux directions

see the peaked density function pulsate slightly along the Φ-axis with period τ_d, as the particles of the ensemble occupy slightly different guiding drift shells with slightly different Φ-values between integer multiples of the drift period (this is the physical meaning of relation (A.74)). In real space, on the other hand, the initial drift shell may change appreciably as the field changes (so as to maintain the enclosed magnetic flux constant); but it, too, will be "sharp" (i.e., occupied by *all* guiding centers) only at integer multiples of τ_d. When, on the other hand, the field variations do not fulfill the adiabatic conditions and do change on a time-scale shorter than τ_d, the point representing the initial group of particles will wander around in invariant space and the delta function singularity will dissolve irreversibly (see next section).

We return to the general case and take a directional differential particle detector to a point $\boldsymbol{r}_0$ of any of the drift shells of Fig. 4.20 with Φ-value in the specified range, pointing it in the direction given by the particles' *local* μ-value. The directional flux $j(\boldsymbol{r}, T, \mu)$ (expression (4.13)) with μ and T to be obtained from M and J, Eqs. (3.14)–(3.17) will be related to the invariant space density by

$$j(\boldsymbol{r}, T, \mu) = \frac{p^2}{2\pi} f_p(\boldsymbol{p}) = \frac{p^2}{2\pi m} F(M, J, \Phi) = \frac{T}{\pi} F(M, J, \Phi) \qquad (4.37)$$

The last equality is valid only in the non-relativistic case. It will not hurt to point out again the following steps to be taken, to relate F to j: Given the shell $[M, J, \Phi]$, (i) pick one point $\boldsymbol{r}_0$ on that shell (Fig. 4.20). (ii) Consider the local B and E (or scalar potential V of the local field line). (iii) Using J (or K) determine the local B_m and μ. (iv) Using M determine T. In the figure, the mirror point traces are the contours of B_m points, where the mirror field intensity must be determined for each field line in question (unless there are no electric fields).

Under special circumstances, other invariant coordinates can be more practical. For instance, we may introduce the triplet M, K (3.14) and L^* (3.41), in which only M depends on the particle energy and in which L^* is a more intuitive shell coordinate than Φ. However, there is a limitation to the use of this set of coordinates: K is an invariant only if field lines are *equipotentials*, i.e., if there are no field-aligned electric fields (Sect. 3.2). Particle shells will now be totally specified by only two field-geometric quantities: L^* and K. When there are no electric fields at all, the set T, B_m, L^* can be used (the good old B-L space); under this restriction, each one is a particle invariant as long as there are no time variations at all.

At this stage, it is convenient to recall the meaning of L^* in asymmetric fields (Sect. 3.5): it represents the radial parameter of *that* shell (its McIlwain L-value) on which a particle would end up if the real field were adiabatically transformed into a *dipole reference field*, with all other forces (the electric field) adiabatically turned off. The fact that, in the real field, particles at a given longitude with the same L^* value are on different field lines (Fig. 3.24), while particles on the same field line have different L^* values, should not confuse the image of L^* as a parameter ordering particles radially: indeed, we should always mentally make the transformation to the reference dipole field when dealing with L^*. In what follows we shall drop the asterisk in L^*. We also should recall that in that transformation the values of a particle's kinetic energy T^* in that dipole reference field will be different and so will be the flux j^*. Expression (4.37) can be used to obtain the following relation:

$$j^* = (T^*/T)j \tag{4.38}$$

For each one of the above examples of coordinates, we'll have a distribution function, $F(M, J, \Phi)$, $F_L(M, J, L)$ and $F_K(M, K, L)$ respectively, which will be related among each other (4.9). Evaluating the Jacobian (4.8) for each relationship we obtain the following links:

$$F_L(M, J, L) = 2\pi B_E R_E^3 L^{-2} F(M, J, \Phi) \tag{4.39}$$

$$F_K(M, K, L) = 4\sqrt{2}\pi B_E R_E^2 m^{1/2} L^{-2} M^{1/2} F(M, J, \Phi) \tag{4.40}$$

Please note the conditions discussed above for the last distribution function. To the above set, one should add the relation to the particle flux (4.37) to complete the set of particle ensemble relationships.

4.6 Basics of the Diffusion Process of Trapped Particles

During their lifetime, radiation belt particles are subjected to the action of five distinct processes: (i) injection of charged particles into the trapping region of the magnetosphere; (ii) acceleration; (iii) adiabatic trapping; (iv) diffusion; and (v) loss. Both acceleration and loss processes may be coupled to, or the consequences of, diffusion. If this chain operates at a rate in which injection and loss are roughly in equilibrium, the particle flux and spectrum at any given point in the trapping region are constant; this in turn means that the distribution function in invariant space $F(M, J, \Phi)$ (Sect. 4.5) will remain constant. The trapping mechanisms were discussed in detail in Chaps. 1–3. We will not deal with injection and loss *per se*; in the following we will limit ourselves to describing the fundaments of trapped particle diffusion theory.

Let us begin by summarizing in a nutshell what happens in a drift shell, especially as it is germane to the process of diffusion. Under adiabatic conditions (2.1), (2.2) a particle gyrates around the magnetic field, bounces back and forth between mirror points and drifts azimuthally around the Earth (Chaps. 1–3). Adiabatic theory introduces the concept of guiding center (page 2), guiding field line (page 57) and guiding drift shell (page 80) by averaging out the phases of a particle's periodic cyclotron, bounce and drift motions, respectively. A "guiding center particle" is a virtual particle with the same mass and charge as the original "real particle", carrying a magnetic moment that impersonates the magnetic properties of the washed-out cyclotron motion (Chap. 1). At any given time, under stationary field conditions, the GC particle bounces back and forth along the instantaneous guiding field line (Chap. 2) and as it drifts perpendicularly to the magnetic field under time-constant conditions, it continuously changes guiding field lines, thus generating the guiding drift shell (Chap. 3). Associated to a particle's periodic cyclotron, bounce and drift motions, adiabatic theory introduces the quantities M, J, Φ (relations (1.26), (3.4), (3.38)) which are constants of motion within the adiabatic approximation (pages 10, 82, 62, and Appendix A.3). These adiabatic invariants are chosen as the natural coordinates in invariant space (Sect. 4.5). The kinetic energy of the particle and its pitch angle at any given point of the drift shell (e.g., as given by its longitude ϕ and arc distance s to the minimum-B contour) can be retrieved from the values of M and J (see (3.14)–(3.17)).

When a whole ensemble of identical particles is trapped on the same guiding drift shell, it can be described by the differential, directional particle flux $\boldsymbol{j}(\boldsymbol{r}, \mu, T)$ (Sects. 4.1 and 4.5), related to the distribution function F in invariant space by Eq. (4.37). To keep with common practice, we now call this invariant space distribution with the letter f, so the flux relation reads $f = 2\pi m\, j/p^2$. In all this, the assumption is made that there is no "phase bunching" (Fig. 4.9), i.e., that cyclotron, bounce and drift phase variables are random but equiprobable. This threefold randomness at the three different time scales τ_c, τ_b, τ_d (1.23), (2.34) and (3.7) plays a fundamental role in the trapped particle diffusion process (we already hinted this on pages 82 and 186), just as an evenly distributed but random cyclotron and bounce cycling was responsible for the existence of perpendicular and parallel pressures and temperatures—even for monoenergetic ensembles (Sect. 4.3). The key issue here is the following: the effect of any given short-term time variation of the field (shorter scale than the period under consideration) on particles with the *same* coordinates in invariant space will depend on *where exactly* (with what instantaneous phase value) the particle was caught in its periodic motion (see discussion at end of Appendix A.3). The end result is that, as seen in the example of page 108, trapped particles with identical M, J, Φ values represented by *one point* in invariant space will be "torn apart" and mix with other groups—in other words, *diffuse*! By being torn apart, the particles' common invariant coordinate related to the periodic motion in question will "dissolve"—i.e., its invariance *will be violated* and the distribution function—initially a delta function in invariant space—will spread out.

Note that invariance violation may happen in azimuthally asymmetric fields even if the externally driven time variation is perfectly smooth, i.e., continuous. Only if the field variability time scale is considerably *longer* than the particle periodicity being examined, will the corresponding invariant be conserved (see discussion at the end of Appendix A.3 for the case of the third invariant Φ). This can be re-formulated by saying that as long as all particles on different points of a common cyclotron, bounce or drift orbit are affected nearly equally by an external field variation, they will, *on the average*, remain together on a common cyclic orbit as time goes on (e.g., pages 82 and 186). But when the field change is appreciable during one cycle, we can anticipate that the effect on particles that were initially on a common orbit with different phases will depend on the *autocovariance* in time of the perturbing signal along the orbit during an orbital period τ. If the autocovariance is zero or near-zero (as happens with signals that change very little during τ), the particles will assemble again on a common cyclotron orbit, guiding field line or guiding drift shell, as we demonstrated in Appendix A.3 for the third invariant; this game will repeat itself, and the corresponding invariant will be conserved on the average. If on the other hand there is high autocovariance, as happens with a signal that is in synchrony with τ, systematically affecting different particles on that common orbit differently, the dispersion will be large, and the corresponding invariant will be violated.

It is important to be clearly aware of the dependence of the three relevant time scales τ_c, τ_b, τ_d on the trapping field and the original particle's energy or velocity. If we turn to relations (1.23), (2.34) and (3.7), or the equivalent relations for near-equatorial particles (see (1.52), (2.41) and (3.12), and remember that even 90° equatorial particles *do* have a bounce eigenfrequency!) we obtain for the corresponding equatorial angular frequencies to zeroth (dipole) order:

$$\omega_c = \frac{qB_E}{mL^3}; \quad \omega_b = \frac{3v}{\sqrt{2}LR_E}; \quad \omega_d = \frac{3mv^2L}{2qB_E R_E^2} \tag{4.41}$$

Let us emphasize the following proportionalities (valid also for $\mu \neq 0$):

$$\omega_c \sim (q/m)^1\, v^0\, L^{-3};\quad \omega_b \sim (q/m)^0\, v^1\, L^{-1};\quad \omega_d \sim (q/m)^{-1}\, v^2\, L \quad (4.42)$$

For the relativistic case, instead of m use $m_0\gamma$ and instead of v insert βc. The highest frequency parameter ω_c depends on the local magnetic field intensity (or L) but is independent of the particle energy (velocity); it only begins to decrease at relativistic energies (the "synchrotron effect"). The bounce frequency is proportional to the particle's velocity but does not depend on particle mass; it decreases with increasing L. The drift frequency, the lowest frequency of all three, increases with the square of the velocity (linearly with kinetic energy), and linearly with m and L. The important fact is that as the particle energy increases the gaps between the three frequencies (which for low energies are of the order of a factor of 100) decrease. The same happens as one moves out to larger L-values. This is important for resonance effects: at higher energies, there can be simultaneous resonance with, for instance, the bounce and drift frequencies at frequencies $m\omega_b + n\omega_d$, where m and n are small integer numbers.

Thus far we have considered the randomness in the phases of the periodic motions of trapped particles. To this we must add the possible randomness of the external perturbations per se. Typical perturbations of the fields include the effects of irregular solar wind behavior such as sudden compressions of the whole magnetosphere (see comment on page 69) and sudden increases of the convection dawn-dusk electric field (page 34); stochastic pitch angle changes due to collisions with ionospheric ions; and the appearance of different types of waves triggered in the omnipresent magnetospheric plasma such as large-scale hydromagnetic ultra-low frequency (ULF) waves (e.g., [1]) and natural and anthropogenic electromagnetic very low frequency (VLF) waves (e.g., [2]). These external irregular or periodic disturbances with their own characteristic randomness of course complicate considerably the theoretical treatment of diffusion.

Summarizing: if the time scale δt of the perturbation satisfies $\tau_d \gtrsim \delta t \gg \tau_b \gg \tau_c$, the third invariant will be violated for a given class of particles whilst the first two invariants will be conserved. Typically this occurs for δt of the order of minutes. If $\tau_b \gtrsim \delta t \gg \tau_c$ ($\delta t \approx$ seconds), both Φ and J will be violated; the first invariant is conserved and the guiding center approximation is still useful. Finally, if $\delta t \lesssim \tau_c$ (milliseconds), all three invariants will be violated. It is important to realize that, in general, once a perturbation process has ceased, even if only temporarily, the particles will resume their adiabatic motion—provided the adiabatic conditions still apply. If all three adiabatic invariants break down, the guiding center approximation becomes invalid and we must deal with the full, exact, particle motion (1.1).

Not only externally triggered perturbations can lead to such a breakdown: if a trapped particle runs into a perfectly static field region such as a plasma boundary where the magnetic field changes direction on a small spatial scale, or into a neutral line or sheet where $B \to 0$, the guiding center approximation will suddenly no longer be valid for that particle—a situation which we may visualize as a (virtual)

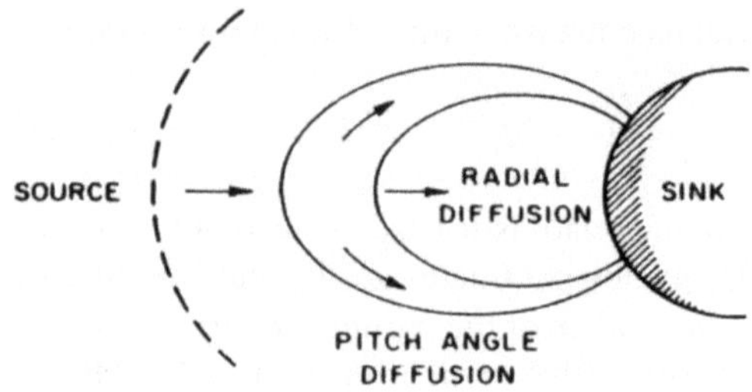

Fig. 4.22 General trapped particle transport by radial and pitch angle diffusion

guiding center particle "decaying" into its original (real) particle. The motion will no longer be adiabatic; for an ensemble of particles it would become randomly thermal, an effect that, too, can be viewed as diffusion.[4]

Diffusion processes of radiation belt particles can be grouped morphologically into two main categories: radial diffusion, which drives particles across drift shells, and pitch angle diffusion, which disperses particle mirror points along field lines (Fig. 4.22). Both kinds usually appear linked intimately with *acceleration mechanisms*. For instance, consider some global change of the trapping magnetic field configuration, of a time scale of the order of a drift period τ_d, leading to a change in the third invariant. As we have shown in Appendix A.3, particularly Fig. A.9 and when going from Eqs. (A.71) to (A.73), during this process the trapped particle changes its bounce-average kinetic energy through the two distinct mechanisms originally described on page 45: a betatron acceleration due to local change in B and a Fermi acceleration due to the effect of the large scale induced electric field along the drift contour. Looking carefully at Fig. A.9 and taking into account the direction of the vector potential, these changes are always such that an *increase* in Φ (decrease in L) will always lead to an *increase* in energy. Therefore, for a general trapped particle source at high L, the resulting inward diffusion through magnetically caused third invariant violation will lead to a general energy increase of the particle population. For many years this *diffusive acceleration* was considered to be a fundamental mechanism for populating and energizing the outer radiation belt (e.g., [3]). The most effective magnetic field variations responsible for this mechanism form a subgroup of the above-mentioned ULF waves—mostly standing Alfvén waves (to be briefly discussed in Sect. 5.5) in the milliHertz range

[4]The much talked-about *magnetic merging mechanism*, a plasma process in which magnetic energy is converted into particle thermal and bulk kinetic energy, can be understood in the above terms of a "guiding center decay": equivalent currents in a plasma (to be discussed in Sect. 5.2) normally sustained by the organized adiabatic behavior of plasma particles and which hold the balance of magnetic field pressure via a macroscopic Lorentz force, break down and the magnetic energy thus freed is converted via induced or polarization electric fields into disorganized thermal particle energy. It should be clear that in this process, part of the ensuing random thermal nature of particle motion is due to the unpredictable cyclotron phase with which a particle is caught by the adiabatic breakdown.

triggered by solar wind pressure fluctuations. Their power spectrum covers the resonance frequencies of drift (and also bounce) periods of particles trapped in the outer belt (e.g., [4,5]), with electric field vectors that may be polarized in the particle drift direction (*poloidal* waves) with maximum Fermi energization if in appropriate resonance with the drift periodicity. This mechanism even works in azimuthally symmetric fields, i.e., at low L values (e.g., [6, 7]). The *toroidal* ULF waves exhibit a radial electric field polarization; they cannot directly energize the particles, but in strongly compressed (azimuthally asymmetric) drift orbits such as at higher L values during a compressed state of the magnetosphere, there will always be a small radial component of the bounce-average drift velocity (examine the equatorial orbit equation (3.20)), and acceleration is possible (e.g., [8]).

The violations of the first and second adiabatic invariants mainly manifest themselves as changes in pitch angle and may or may not be coupled to energy changes. The physically simplest mechanism is pitch angle scattering by elastic Coulomb interactions with atoms of the ionosphere. Turning to the expression (1.26) of the magnetic moment, such interactions change M without affecting kinetic energy and guiding field line (within a gyroradius): particles diffuse in M, B_m or μ space along their field lines and, considering Fig. 3.13, end up running into the bounce loss cone, i.e., precipitating into the atmosphere (unless quickly replenished along longitudinal drift—see the old "windshield-wiper effect" on page 76), or they end up diffusing into the magnetopause (drift loss cone). A more complex, but more general, pitch angle diffusion mechanism consists of resonant interactions with VLF waves, which in essence are magnetically guided electromagnetic waves whose $\boldsymbol{E}$-vector, if circularly polarized, may enter in resonance with the (appropriately Doppler-shifted) cyclotron frequency of trapped particles. When these VLF waves are coherent (the so-called chorus emissions, found outside the plasmapause mostly during times of sunward-convecting tail plasma—see regions outside the separatrices in Fig. 1.21), cyclotron resonance can lead to both, pitch angle scattering and acceleration. In the latter case we are in presence of a *local acceleration* mechanism, spatially delimited, as opposed to the radial diffusion-driven acceleration mentioned above. There are also incoherent VLF waves (the so-called plasmaspheric hiss, trapped in a limited region near the plasmapause), which do not energize but pitch-angle-scatter the particles. They are believed to be responsible for the formation of the "slot" separating the inner radiation belt from the outer one. The recent unexpected finding of a newly-formed "third electron Van Allen belt" precisely inside that slot [9], points to local acceleration as a fundamental, perhaps the most important, trapped particle energization mechanism. Most of the preceding description refers to trapped electrons. Ions (protons, alpha particles) are also subjected to these mechanisms, but there is a host of other magnetospheric plasma waves (e.g., electromagnetic ion cyclotron waves, electrostatic electron cyclotron harmonic waves) with specific effects on these particles [2].

Diffusion theory is used to calculate the time evolution of the particle distribution function. The usual recipe is as follows: (i) Choose an adequate invariant 3-D coordinate space in which the whole particle population is to be described by a distribution function. (ii) Set up a diffusion equation (see next section). (iii) Single out the individual diffusion mechanisms that will be taken into account, and establish mathematical expressions for the corresponding diffusion coefficients, identifying adjustable parameters related to the models of the fields, their variations, and the postulated interaction mechanisms. (iv) Define the source and loss mechanisms. (v) Integrate the diffusion equation (numerically, or analytically for near-equatorial particles). (vi) Adjust the parameters so as to achieve a best fit to the data set at hand. Sounds straightforward, but almost as a rule there are several different mechanisms coupled together, which makes finding the coefficient expressions difficult and the mathematical description multidimensional. For a basic detailed treatment, see the classic [3]; for a recent summary, see [2].

In the following sections we will discuss violations of the adiabatic invariants, with a discussion of the governing evolution equation. We shall focus on the third adiabatic invariant since it is the most likely to be violated by large solar wind triggered perturbations. Since the theory of radiation belt diffusion is still evolving (e.g., [10–13])[5] we shall discuss only some main guidelines with some explanation of the physics involved. A detailed analysis of diffusion theory and recent experiment would require several chapters, enough to fill a separate book.

4.7 Derivation of the Fokker-Planck Equation

Let us consider a time-dependent distribution function of trapped particles $f_0(\Phi, J, M, t)$. We should remember that this is the distribution function of the particles' guiding centers (designated $F(M, J, \Phi)$ in Sect. 4.5) that is proportional to the particle density in phase space and, hence, to the *measurable quantity* j/p^2 (4.13) (proportional to j/T non-relativistically). To derive an evolution equation for f_0 we must relate this distribution function at time $t + \Delta t$ with the distribution function of the same group of particles at time t. We shall set $\Delta t \simeq \tau_d$; furthermore, we assume that for stochastic interactions during Δt there are *very many* individual acts of random changes of the variable in question, yet still with a small total effect.

To facilitate the derivation of the Fokker-Planck equation, let us initially limit the case to just *one variable*, say, the third invariant Φ: $f_0 = f_0(\Phi, t)$. As the first

[5]Historically, trapped particle diffusion theory began with gusto during the early days of radiation belt study, stimulated by an ideal experimental situation (and Cold War military encouragement): several high-altitude nuclear detonations were conducted by the USA and the USSR between 1958 and 1962, in which large fluxes of trapped electrons from bomb fission products were injected into a limited range of L-shells. The time evolution of these trapped electron fluxes could be measured with radiation detectors on early satellites under most ideal conditions: a well known initial state in time, space and energy spectrum and a greatly limited initial trapping region! No such controlled conditions occurred afterwards; only natural events like big geomagnetic storms provided some identifiable initial conditions.

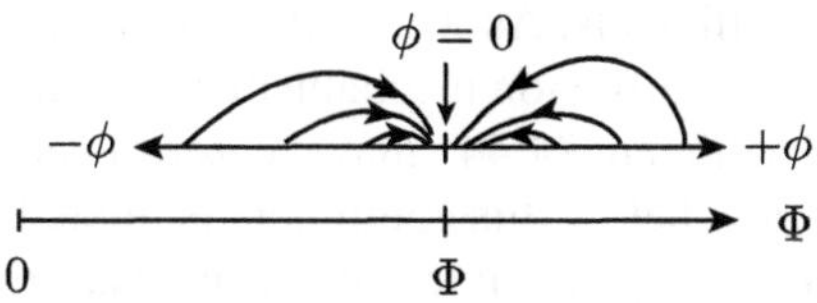

Fig. 4.23 Third invariant variables in the definition of the probability function Π

step, we introduce a *probability function* playing a central role in diffusion theory and quantitatively representing the physical mechanism responsible for the small random variations of the third invariant Φ. We designate it $\Pi(\Phi - \phi, \phi, t)$, the probability that a particle with an invariant shell coordinate $\Phi - \phi$ at time t will end up with coordinate Φ at time $t + \Delta t$.

Figure 4.23 sketches the variables involved in this process. The two horizontal axes represent the one-dimensional invariant space of coordinate Φ, and the overlapping domain of the change ϕ in third invariant of a particle after the time lapse Δt (the variable of the integrations below). The sketched arcs indicate possible "jumps" of particles' Φ-values from their state at time t to the Φ-value at time $t + \Delta t$ (each jump really made up of many small ones).

For a given perturbation, Π represents not only the effect on particles due to random external physical effects (induced electric field accelerations, collisions) but also the effect due to the fact that in an asymmetric B-field, under equal external perturbations, the actual change of Φ will depend on the instantaneous position (phases) of the particle on the drift shell (e.g., its longitude and latitude) (see comments in Appendix A.3 concerning Fig. A.9). If Δt is of the order of τ_d, $\Pi(\Phi - \phi, \phi, t)$ will be proportional to Δt. As any legitimate probability function, Π must be normalized:

$$\int_{\Phi-space} \Pi d\phi = 1 \tag{4.43}$$

In principle, during the interval Δt particles may jump into Φ from any third invariant value, so the integral must be taken over the entire Φ-space.

The quantity $\int_{\Phi} \phi \Pi d\phi = \langle \Delta\Phi \rangle$ is the average change of Φ for one particle on the shell Φ during that time interval. The average change *per unit time* is called the first order Fokker-Planck *diffusion coefficient*:

$$D_{\Phi} = \frac{\langle \Delta\Phi \rangle}{\Delta t} = \frac{1}{\Delta t} \int_{\Phi-space} \phi \Pi d\phi \tag{4.44}$$

Likewise, we have a second order coefficient

$$D_{\Phi\Phi} = \frac{\langle \Delta(\Phi)^2 \rangle}{\Delta t} = \frac{1}{\Delta t} \int_{\Phi-space} \phi^2 \Pi d\phi \tag{4.45}$$

In the case of several invariant variables (see below), integrals of the product of any two of them (always weighted with Π) will also be second order diffusion coefficients. As we shall see below, second order coefficients represent the *strength* of the diffusion mechanism; when they are zero, there is no diffusion. The first order coefficient (4.44) may be zero even if diffusion is present; however, as we shall show below this coefficient is related to the corresponding second order coefficient. If during the interval Δt no variations of Φ are expected, i.e., if the invariant coordinate remains constant (this does not mean that no changes are taking place, only that they are adiabatic, on a time-scale much longer than τ_d), the probability function is a delta-function, $\Pi = \delta(\phi)$ ($\phi = 0$ means no change), and the corresponding diffusion coefficients will be zero.

To find the evolution equation for f_0 we relate the distribution function at time t with that at time $t + \Delta t$:

$$f_0(\Phi, t + \Delta t) = \int_{\Phi-space} f_0(\Phi - \phi, t)\, \Pi(\Phi - \phi, \phi, t) d\phi + \big(Q(\Phi, t) - S(\Phi, t)\big)\Delta t$$

We added the functions Q and S to represent source injections and sink absorptions per unit time. Expanding in Taylor series both f_0 and Π up to the second order in Φ in the integral, and f_0 to first order in t on the left side, sorting and grouping terms, and taking into account (4.43) and the definition of diffusion coefficients (4.44) and (4.45), we obtain the one-dimensional *Fokker-Planck equation*

$$\frac{\partial f_0}{\partial t} = -\frac{\partial}{\partial \Phi}(D_\Phi f_0) + \frac{1}{2}\frac{\partial^2}{\partial \Phi^2}(D_{\Phi\Phi} f_0) + Q - S \tag{4.46}$$

We now introduce a quantity which we may call the *net diffusion velocity*. It is the average *collective* displacement per unit time in invariant space of an ensemble of trapped particles due to diffusion (in the present one-dimensional case, the average rate of change of Φ of the ensemble of particles, that at time t were in a given $\delta\Phi$ bin). If we designate this velocity by $\mathring{\Phi}$, the conservation of number of particles (continuity equation in one-dimensional Φ-space) would be written as

$$\frac{\partial f_0}{\partial t} = -\frac{\partial}{\partial \Phi}(\mathring{\Phi} f_0) \tag{4.47}$$

Combining with Eq. (4.46) and integrating over Φ, we obtain (assuming there are no sources nor sinks):

$$\mathring{\Phi} = D_\Phi - \frac{1}{2 f_0}\frac{\partial(D_{\Phi\Phi} f_0)}{\partial \Phi} = D_\Phi - \frac{1}{2}\frac{\partial D_{\Phi\Phi}}{\partial \Phi} - \frac{D_{\Phi\Phi}}{2 f_0}\frac{\partial f_0}{\partial \Phi} \tag{4.48}$$

The integration constant was set to zero. Each term in the above equation has a specific physical meaning. We begin with the third term containing the gradient of f_0 in Φ-space and take a look at Fig. 4.23. If initially there are more particles on the side of positive ϕ values (i.e., a positive gradient of f_0 at $\phi = 0$, more particles with a third invariant value greater than the Φ to which they will jump), the end effect of diffusion after the time interval Δt will be a general migration of particles to the *left* in the figure, indicating a negative net diffusion velocity. Clearly, the coefficient $D_{\Phi\Phi}$ represents the strength of the diffusive process. Now let us consider the second term in (4.48). A positive gradient of $D_{\Phi\Phi}$ means a diffusion process more effective on the right side than on the left, which again will lead to a higher probability of particles jumping to the left during Δt. Finally, regarding the first term, it represents, by the very definition of the first order diffusion coefficient D_{Φ} (4.44), the measure of unequal dispersion of particles from their initial position in Φ-space; if D_{Φ} is positive, it represents a positive contribution to the net velocity $\overset{\circ}{\Phi}$.

This discussion provides the opportunity for a "kindergarten demonstration" of an important relation between the first and second diffusion coefficients. Consider the case of a *uniform* distribution function f_0 = const. According to (4.48), the net diffusion velocity would be $\overset{\circ}{\Phi} = D_{\Phi} - 1/2(\partial D_{\Phi\Phi}/\partial\Phi)$, independent of the actual value of the distribution function. Given the meaning of net diffusion velocity, the value of $\overset{\circ}{\Phi}$ must be zero for a uniform distribution: whatever group of particles moves away from their initial Φ bin, should be replaced by an equal number of particles coming into that bin during Δt. This can happen only if at all times the contribution of D_{ϕ} exactly cancels that of $1/2(\partial D_{\Phi\Phi}/\partial\Phi)$:

$$D_{\Phi} = 1/2\left(\frac{\partial D_{\Phi\Phi}}{\partial\Phi}\right) \tag{4.49}$$

This relationship leaves us with the following simpler expressions for the one-dimensional Fokker Planck equation and net diffusion velocity:

$$\frac{\partial f_0}{\partial t} = \frac{1}{2}\frac{\partial}{\partial\Phi}\left(D_{\Phi\Phi}\frac{\partial}{\partial\Phi}f_0\right) + Q - S \tag{4.50}$$

$$\overset{\circ}{\Phi} = -\frac{D_{\Phi\Phi}}{2f_0}\frac{\partial f_0}{\partial\Phi} = -\frac{D_{\Phi\Phi}}{2}\frac{\partial\ln(f_0)}{\partial\Phi} = -\frac{D_{\Phi\Phi}}{2}\frac{\partial}{\partial\Phi}\left(\ln\frac{j}{T}\right) \tag{4.51}$$

Taking into account (4.37), this clearly shows that diffusive flows run from higher to lower phase space density.

The general one-dimensional equation (4.46) can be extended easily to the *three invariant variables* M, J, Φ to obtain:

$$\begin{aligned}\frac{\partial f_0}{\partial t} = & -\frac{\partial}{\partial\Phi}(D_{\Phi} f_0) - \frac{\partial}{\partial J}(D_J f_0) - \frac{\partial}{\partial M}(D_M f_0)\\ & +\frac{1}{2}\frac{\partial^2}{\partial\Phi^2}(D_{\Phi\Phi} f_0) + \frac{1}{2}\frac{\partial^2}{\partial J^2}(D_{JJ} f_0) + \frac{1}{2}\frac{\partial^2}{\partial M^2}(D_{MM} f_0)\\ & +\frac{\partial^2}{\partial\Phi\partial J}(D_{\Phi J} f_0)\frac{\partial^2}{\partial\Phi\partial M}(D_{\Phi M} f_0) + \cdots + Q - S\end{aligned} \tag{4.52}$$

The probability function must be normalized in the 3-D invariant space:

$$\iiint \Pi \, d\phi \, dJ \, dM = 1 \tag{4.53}$$

The diffusion coefficients in (4.52) are

$$\begin{aligned}
D_\Phi &= \frac{1}{\Delta t} \int \phi \Pi \, d\phi = \frac{\langle \Delta \Phi \rangle}{\Delta t} \\
D_{\Phi\Phi} &= \frac{1}{\Delta t} \int \phi^2 \Pi \, d\phi = \frac{\langle \Delta \Phi^2 \rangle}{\Delta t} \\
D_J &= \frac{1}{\Delta t} \int \eta \Pi \, d\eta = \frac{\langle \Delta J \rangle}{\Delta t} \\
&\cdots\cdots\cdots\cdots \\
D_{\Phi J} &= \frac{1}{\Delta t} \int \int \phi \eta \Pi \, d\phi d\eta = \frac{\langle \Delta \Phi \Delta J \rangle}{\Delta t} \\
&\cdots\cdots\cdots\cdots
\end{aligned} \tag{4.54}$$

The same criteria as for the elimination of the first order diffusion coefficient D_Φ can be applied to D_M and D_J, which leads to a simplified equation. Uniformizing the notation by changing the three natural invariants M, J, Φ to X_1, X_2, X_3 and introducing diffusion coefficients D_{X_i,X_j} that are *one-half* of the original Fokker-Planck coefficients (4.54), we can write:

$$\frac{\partial f_0}{\partial t} = \sum_{i,j=1}^{3} \frac{\partial}{\partial X_i} \left(D_{X_i X_j} \frac{\partial f_0}{\partial X_j} \right) \tag{4.55}$$

If for practical reasons it is advisable to switch to another system of invariant coordinates, one must use the rules of transformation discussed in Sects. 4.2 and 4.5. For an arbitrary coordinate system (I_1, I_2, I_3), the Fokker Planck equation will be:

$$\frac{\partial}{\partial t} f(I_1, I_2, I_3, t) = \sum_{i,j=1}^{3} \frac{1}{G_{XI}} \frac{\partial}{\partial I_i} \left(G_{XI} \tilde{D}_{I_i I_j} \frac{\partial f}{\partial I_j} \right) \tag{4.56}$$

where G_{XI} is the Jacobian (4.8) for the transformation from (X_1, X_2, X_3) to (I_1, I_2, I_3). $\tilde{D}_{I_i I_j}$ are the new diffusion coefficients in the (I_1, I_2, I_3) coordinate system, which can be calculated from the original diffusion coefficients $D_{X_i X_j}$ by

$$\tilde{D}_{I_i I_j} = \sum_{k,l=1}^{3} \frac{\partial I_i}{\partial X_k} D_{X_k X_l} \frac{\partial I_j}{\partial X_l} \tag{4.57}$$

As an example, we consider again the case in which there is no other diffusion except one caused by violations of the third invariant. We choose the more convenient distribution function $F(M, J, L)$ of Sect. 4.5, the corresponding Jacobian to relate the old variables to the new ones has only one element, derived from (3.41): $G_{\Phi L} = |\partial L/\partial \Phi| = L^2/2\pi B_E R_E^2$. Equation (4.50) becomes

$$\frac{\partial f(M,J,L)}{\partial t} = \frac{\partial}{\partial L}\left[D_{LL}\frac{1}{L^2}\frac{\partial}{\partial L}(L^2 f(M,J,L))\right] + Q - S \qquad (4.58)$$

Concerning the net diffusion velocity in this case, we obtain

$$\mathring{L} = -D_{LL}\frac{\partial}{\partial L}\left[\ln\left(\frac{j}{T}\right)\right] \simeq -D_{LL}\frac{\partial}{\partial L}\left[\ln\left(jL^3\right)\right] \qquad (4.59)$$

The last near-equality considers that due to conservation of the first invariant M, the kinetic energy of a radially diffusing particle varies like $T \sim B \sim L^{-3}$. Note that it is the gradient of jL^3 that determines the net radial diffusion velocity. This is why a plot of jL^3 vs L helps identify the regions in the magnetosphere where radial diffusion is inwards or outwards. Concerning the radial diffusion coefficient D_{LL}, if the process is in any way dependent on fluctuations of the azimuthal magnetic field asymmetry and its effect on drift shells (3.42), it should be proportional to the drift-average of the fluctuations of σ_A^2 (see (3.43) and (1.50)):

$$D_{LL} = \langle(\Delta L)^2\rangle/\tau_d \sim \langle(\Delta\sigma)^2\rangle \sim R_s^{-8}L^{10}(\Delta R_s/R_s)^2$$

where $\Delta R_s/R_s$ is the relative fluctuation of dipole-distortion parameter. Note the strong dependence on the radial parameter L and also on R_s which in the simple model that was used to derive relation (3.45) is the stand-off distance R_s (but in reality would be represented by a parameter globally describing magnetospheric activity, such as K_p or, in the case of ULF waves, the wave power). For an azimuthally symmetric magnetic field under azimuthally symmetric perturbations, the coefficient b_2 in (1.50) is zero and according to (3.43), σ_A^2 would remain zero. Under these circumstances, radial diffusion would not be possible: shells would fluctuate, but according to our comments on page 186 of Appendix A.3, retain at all times the value of their third invariant, i.e., their L-value. As a consequence, the trapped particle population in a pure dipole with stochastically fluctuating dipole strength (or a fluctuating symmetric compression as described by the parameter b_1) would "jitter or pulsate" radially, but never diffuse. Only a longitudinally localized perturbation, like an azimuthal electric field parallel to the particles' drift velocity would place them on drift shells with different Φ or L values.

Fokker-Planck diffusion theory is "the art of creating pleasing diffusion coefficients" (where "pleasing" means yielding solutions of the diffusion equation that are in agreement with the data). Among other things, concerning large scale third and second invariant violations, it is always the *induced electric field* $\boldsymbol{E}_{ind} = -\partial A/\partial t$ which ultimately is responsible for orbital and energy changes of a trapped particle, for any given time variation of $\boldsymbol{B}$, whether big or small, long-term or short-term.

This demands skill in handling correctly the vector potential $\boldsymbol{A}$ (see pages 83 and 176 in Appendix A.1 and A.3). Concerning the smaller scale first invariant violations, the key physical quantity is the wave electric field, which in turn requires great skill in handling the bewildering host of waves in the magnetospheric plasma. Currently, there is a rapidly increasing activity in experimental and theoretical research on radiation belt diffusion (e.g., [12, 14] respectively). Still, as of now, a compilation of diffusion coefficients shows differences of many orders of magnitude between the values obtained or used by different authors. It would fall outside of the scope of this book to go into details.

References

1. M.G. Kivelson, Pulsations and magnetohydrodynamic waves, in *Introduction to Space Physics*, ed. by M.G. Kivelson, C.T. Russell (Cambridge University Press, Cambridge, 1995), pp. 330–355
2. R.M. Thorne, Radiation belt dynamics: the importance of wave-particle interactions. Geophys. Res. Lett. **37**, 22107 (2010)
3. M. Schulz, L.J. Lanzerotti, *Particle Diffusion in the Radiation Belts* (Springer, Berlin, 1974)
4. D.J. Southwood, M.G. Kivelson, Charged particle behavior in low-frequency geomagnetic pulsations. I. Transverse waves. J. Geophys. Res. **86**, 5643–5655 (1981)
5. D.J. Southwood, M.G. Kivelson, Charged particle behavior in low-frequency geomagnetic pulsations. II – graphical approach. J. Geophys. Res. **87**, 1707–1710 (1982)
6. Q.-G. Zong, X.-Z. Zhou, X. Li, P. Song, S.Y. Fu, D.N. Baker, Z.Y. Pu, T.A. Fritz, P. Daly, A. Balogh, H. Réme, Ultralow frequency modulation of energetic particles in the dayside magnetosphere. Geophys. Res. Lett. **34**, 12105 (2007)
7. Q.-G. Zong, X.-Z. Zhou, Y.F. Wang, X. Li, P. Song, D.N. Baker, T.A. Fritz, P.W. Daly, M. Dunlop, A. Pedersen, Energetic electron response to ULF waves induced by interplanetary shocks in the outer radiation belt. J. Geophys. Res. **114**, 10204 (2009)
8. S.R. Elkington, M.K. Hudson, A.A. Chan, Resonant acceleration and diffusion of outer zone electrons in an asymmetric geomagnetic field. J. Geophys. Res. **108**, 11–1 (2003)
9. D.N. Baker, S.G. Kanekal, V.C. Hoxie, M.G. Henderson, X. Li, H.E. Spence, S.R. Elkington, R.H.W. Friedel, J. Goldstein, M.K. Hudson, G.D. Reeves, R.M. Thorne, C.A. Kletzing, S.G. Claudepierre, A long-lived relativistic electron storage ring embedded in Earth's outer Van Allen belt. Science **340**(6129), 186–190 (2013)
10. I.R. Mann, K.R. Murphy, L.G. Ozeke, I.J. Rae, D.K. Milling, A. Kale, F. Honary, The role of ultralow frequency waves in radiation belt dynamics, in *Dynamics of the Earth's Radiation Belts and Inner Magnetosphere*, vol. 199, ed. by D. Summers et al., Geophysical Monograph Series (AGU, Washington, DC, 2012), pp. 69–91
11. R.A. Treumann, W. Baumjohann, *Advanced Space Plasma Physics* (Imperial College Press, London, 1997)
12. G.D. Reeves, H.E. Spence, M.G. Henderson, S.K. Morley, H.W. Friedel, H.O. Funsten, D.N. Baker, S.G. Kanekal, J.B. Blake, J.F. Fennell, S.G. Claudepierre, R.M. Thorne, D.L. Turner, C.A. Kletzing, W.S. Kurth, B.A. Larsen, J.T. Niehof, Electron acceleration in the heart of the Van Allen radiation belts. Science **341**(6149), 991–994 (2013)
13. S. Lejosne, D. Boscher, V. Maget, G. Rolland, Deriving electromagnetic radial diffusion coefficients of radiation belt equatorial particles for different levels of magnetic activity based on magnetic field measurements at geostationary orbit. J. Geophys. Res. **118**, 3147–3156 (2013)
14. S. Lejosne, Modélisation du phénomène de diffusion radiale au sein des ceintures de radiation terrestres par technique de changement d'échelle. PhD thesis, Université de Toulouse, France, 2013

Chapter 5
Collisionless Plasmas

5.1 Introduction: From Individual Particles to Fluids

In the previous chapters we have studied the dynamics of trapped particles in given magnetic and electric fields of sources that are *external* to the particle population of interest (such as the geomagnetic field, boundary and cross-tail currents, solar-wind-imposed electric field, etc.). On occasion we examined the electric currents and charge density generated by the particles as the result of their motion in the given fields, but neglected any retro-effects on such fields and resulting feedback on the collective behavior of the particles that caused such effects in the first place.

The population of radiation belt particles represents a very small proportion of total kinetic energy and mass in the terrestrial magnetosphere. The magnetosphere itself is shaped by currents carried by a much denser, quasi-neutral and mostly collisionless ensemble of lower energy ions and electrons confined by the magnetic field—the *magnetospheric plasma*—representing the bulk of kinetic energy flow and mass[1] in the terrestrial outer environment. In the remaining sections of this book we shall study the dynamics of these multi-species particle ensembles as a natural extension of our preceding discussion of the adiabatic theory of individual particle motion. This approach (sometimes called orbit-theory approach) will allow us to gain a better physical understanding of the sometimes intricate structure and behavior of a plasma as represented by local macroscopic temporal and/or spatial averages of particle properties—the *macroscopic fluid variables* for which our instruments provide the data and which we invoke whenever we picture mentally, describe mathematically or model numerically a collisionless plasma. Our ultimate aim is to analyze and help understand a plasma and its electromagnetic fields as one whole—a self-organizing entity with distinct but thoroughly interacting regions which in general cannot be studied and understood in isolation from each other.

---

[1]This sounds a bit pompous. The maximum total energy flow in the astronomically-sized magnetosphere can be estimated at barely $\sim$20,000 MW, its total plasma mass a mere 20 t.

J.G. Roederer and H. Zhang, *Dynamics of Magnetically Trapped Particles*, Astrophysics and Space Science Library 403, DOI 10.1007/978-3-642-41530-2_5,

There are several ways to visualize and represent an ensemble of charged particles in a magnetic field. For pedagogical reasons, at this early stage of our discussion we will make three fundamental simplifying assumptions. First, we shall *consider separately* only one of the different species of charged particles that may constitute a plasma (electrons, protons, alpha particles, heavier ions). Second, in considering only one given species, we shall *neglect any electrostatic effects* caused by the spatial accumulation of same-sign electric charges (for instance, by assuming that in the case of a positive ion ensemble, there would be enough low-energy ambient electrons around to neutralize any such effect). Third, we shall *neglect collisions* and the field singularities in the proximity of each particle (i.e., assume a continuous, finite magnetic and electric field everywhere in the ensemble). In addition, we shall consider only *non-relativistic* particle ensembles (a limitation that excludes plasmas in the extreme environments of neutron stars and black holes). Later we shall turn to the realistic situation of an electrically quasi-neutral *mixture* of at least two different species of opposite charges. In all this, the particle distribution function (see Sect. 4.1 for definition and examples) will be the "workhorse" for the initial mathematical description of an ensemble, providing the link between microscopic properties and more intuitive and measurable macroscopic physical variables. Let us point out that as mentioned in Sect. 4.1, a distribution function already represents an average, in which an enormous number of degrees of freedom (the exact positions and velocities of each one of the particles in an ensemble) are condensed into just six variables—the coordinates of a point in 6-D *phase space* at which the distribution function represents a density (number of particles per unit volume of coordinate and momentum space).

Our point of departure will be the *kinetic theory* of an ensemble of charged particles, each species of which is described by a time-dependent particle distribution function $f(\boldsymbol{r}, \boldsymbol{p}, t)$ in phase space $\{\boldsymbol{r}, \boldsymbol{p}\}$, obeying Liouville's equation (4.25). In non-relativistic plasma physics, it is customary to define the particle distribution function in *velocity* subspace, as we did in Sect. 4.4; henceforth we shall use the general distribution function $f = f(\boldsymbol{r}, \boldsymbol{v}, t)$. Taking into account (1.1) for the local force (and neglecting other external forces such as gravitation), we write the Liouville equation in the form:

$$\frac{\partial f}{\partial t} + \boldsymbol{v} \cdot \nabla_r f + \frac{q}{m}(\boldsymbol{E} + \boldsymbol{v} \times \boldsymbol{B}) \cdot \nabla_v f = 0 \tag{5.1}$$

This is called the *Vlasov equation*, basis of the kinetic theory of collisionless plasmas. The vector operator ∇_v has components $\partial/\partial v_i$.

As shown in Sect. 4.3, the distribution function f serves to define macroscopic quantities as average values of physical variables of the ensemble particles. For a given species, we list them again:

Number density: $n(\boldsymbol{r}, t) = \int f \, d\boldsymbol{v}$
Mass density: $\rho_m(\boldsymbol{r}, t) = mn(\boldsymbol{r}, t)$
Bulk (or average) velocity: $\boldsymbol{V} = \langle \boldsymbol{v} \rangle = \int \boldsymbol{v} f \, d\boldsymbol{v} / \int f \, d\boldsymbol{v}$

Kinetic (or momentum flux) tensor (see (4.15)): $\mathbb{K}(\boldsymbol{r},t) = m\int \boldsymbol{v}\otimes\boldsymbol{v}\, f\, d\boldsymbol{v}$
Pressure tensor (kinetic tensor in a frame traveling with bulk velocity $\boldsymbol{V}$, (4.16)):
$\mathbb{P}(\boldsymbol{r},t) = m\int(\boldsymbol{v}-\boldsymbol{V})\otimes(\boldsymbol{v}-\boldsymbol{V}) f\, d\boldsymbol{v} = \mathbb{K} - mn\,\boldsymbol{V}\otimes\boldsymbol{V}$
Average kinetic energy density: $\epsilon = 1/2\,nm\langle v^2\rangle = 1/2\,m\int v^2 f\, d\boldsymbol{v}$
Internal energy density (kinetic energy density in a frame traveling with bulk velocity $\boldsymbol{V}$): $w = 1/2\,m\int(\boldsymbol{v}-\langle\boldsymbol{v}\rangle)^2 f\, d\boldsymbol{v} = \epsilon - 1/2nmV^2 = 1/2\,\mathrm{Tr}\mathbb{P}$
Charge density: $\rho_q(\boldsymbol{r},t) = q\,n(\boldsymbol{r},t)$
Electric current density: $\boldsymbol{J} = q\,n\boldsymbol{V} = q\int \boldsymbol{v}\, f\, d\boldsymbol{v}$

These are the fundamental macroscopic variables for a *particle fluid* (also called kinetic fluid) description of an ensemble of charged particles.[2] Notice that the electromagnetic field does not appear explicitly, except in the expression of the forces on the particles that ultimately control the distribution function via the Vlasov equation.

5.2 The Guiding Center Fluid Model

In many situations of collisionless magnetospheric plasmas the constituent charged particles behave adiabatically, i.e., they gyrate rapidly in cyclotron motion perpendicular to $\boldsymbol{B}$, they move parallel to $\boldsymbol{B}$ (bounce, if the field geometry is right) and they drift perpendicularly to $\boldsymbol{B}$—as long as the conditions (2.1) and (2.2) hold for the particles and the field. As we have done in the adiabatic theory of single particles in Chaps. 1–3, the mathematical description and mental visualization of the ensemble can be simplified by averaging all dynamic variables over one cyclotron turn and replacing each "madly gyrating" particle by a virtual particle at its guiding center, bouncing along and drifting across magnetic field lines. However, to accomplish

[2] A brief detour into Foundations of Physics is in order here. In the Preface we already stated that "physics is the art of modeling", and in Sect. 1.1 we introduced the model of a "guiding center particle". A fluid (any fluid!) is also a model—the model of a system in which a huge, mathematically unmanageable, number of physically real particles (molecules, atoms, electrons, nucleons, quarks, gluons, sand grains, etc., depending on the ensemble in question) has been replaced in our mental image and in the quantitative description by a virtual *continuum* (see also Appendix A.1, page 160). We speak of and quantitatively describe "parcels" of fluid and imagine how they are deformed as they move, and guided by what our physiological senses experience when exposed to liquids or flowing gases, we introduce macroscopic variables which can be used for practical purposes, like density, bulk velocity, pressure, temperature, internal energy, entropy, etc. Statistical mechanics and, as a corollary, plasma physics were developed to link approximate but intuitive macroscopic continuum descriptions of matter with their physically real microscopic structures that can only be revealed through the use of scientific instruments. In our case, distribution functions and the differential equations which they obey establish such a link. The main aim of any fluid description is to formulate physical-mathematical relationships between the macroscopic variables so as to provide a "coarse-grained" quantitative description of the dynamic state of the ensemble—regardless of the unknowable detailed state (position, velocity) of each elementary constituent.

this we must require as an additional condition that the particle distribution be *gyrotropic* (strictly speaking, gyrotropic in a local coordinate system that moves with the average guiding center velocity (2.14) at that point). This means that there should be no synchronized cyclotron phase bunching (Fig. 4.9); in presence of electromagnetic waves of the order of the particles' gyrofrequency, this assumption is no longer valid. All this allows us to dispense of one degree of freedom, the phase angle φ of a particle's cyclotron motion, and use the distribution function

$$F = F(\boldsymbol{r}, t, v_\perp, v_\parallel) \tag{5.2}$$

where $\boldsymbol{r}$ is not the position $\boldsymbol{r}_p$ of the actual particle but that of its guiding center $\boldsymbol{r} = \boldsymbol{r}_p + (m/q)\boldsymbol{v} \times \boldsymbol{B}/B^2$ (see relation (1.25)). The quantity $\delta n = F(\boldsymbol{r}, t, v_\perp, v_\parallel)\delta\boldsymbol{r}\delta v_\perp \delta v_\parallel$ represents the number of virtual guiding centers in $\delta\boldsymbol{r}$ at point $\boldsymbol{r}$ and time t, whose "parent" particles have velocities $v_\perp$ and $v_\parallel$ in the element $\delta v_\perp \delta v_\parallel$. Of course, it is also possible to use derived distribution functions such as $F(\boldsymbol{r}, t, T, \mu)$ (Sect. 4.2).

Having eliminated φ does not mean that we can neglect collective effects of the cyclotron motion. First, notice the hidden presence of the $\boldsymbol{B}$ vector: at each point in space it defines the $\perp$ and $\parallel$ directions, the natural frame of coordinates (Appendix A.1) (or, in derived distribution functions, the parameters μ, M or I). Second, the magnetic moment $\boldsymbol{M}$ (1.26) generated by the now "washed-out" cyclotron motion must be retained in the contribution of the particles to the magnetic field. Likewise, the particle's angular momentum $\boldsymbol{l} = (2m/q)\boldsymbol{M}$ (1.27) must be retained as a contribution to the macroscopic dynamic state of the fluid. Third, we must retain the contribution of a particle's cyclotron motion to the perpendicular pressure $p_\perp$ and that of its parallel motion to the parallel pressure $p_\parallel$ (4.17), as well as to the internal kinetic energy density w. As a consequence of all this we picture the *guiding center fluid* as a model fluid consisting of *magnetized* virtual GC particles with *intrinsic* angular momentum, and endowed with local vorticity, internal kinetic energy, temperature and perpendicular and parallel pressures. The magnetic field thus assumes in explicit form the role of a "scaffolding", an internal skeleton that greatly aids in visualizing plasmas but whose local asymmetry obliges us to always be aware of the different character of transverse and field aligned properties, respectively.

In particular, concerning the field-aligned motion of the guiding center particles, the conservation of each individual particle's magnetic moment (1.26) provides a fundamental link between points of a given field line in a guiding center fluid. For instance, great care has to be taken with the interpretation and handling of distribution functions in the guiding center fluid model. As shown in Sect. 4.4, they are causally connected along a given field line because of the bounce motion; for instance, in an equilibrium situation in which there is no particle bunching, the distribution function in a guiding center fluid can only be prescribed on a specified *surface* such as the minimum-B surface which is traversed by all trapped particles on a field line (Fig. 4.12); it cannot be chosen arbitrarily all along a field line. In what follows, the position vector $\boldsymbol{r}$ in the distribution function F will usually signify "distribution function at a reference (e.g., minimum-B) point of the field line going through point $\boldsymbol{r}$". Although in the preceding text we have been mentioning bounce motion, in plasma physics neither bounce nor drift periodicities (and the related adiabatic invariants J and Φ) play any direct role; mainly, because the field geometries, their time variations and the presence of multiple but mutually interacting classes of particles do not favor sustained particle trapping. Therefore, for guiding center particles in a GC fluid there will be no bounce-average drift velocities nor any drift-average quantities—only instantaneous ones. The only averaging is done over cyclotron motion.

We can list expressions for the macro-variables in the guiding center description, as we did for a kinetic fluid:

Number density: $n(\boldsymbol{r},t) = \int F\, dv_\perp dv_\parallel$
Mass and charge densities: $\rho_m(\boldsymbol{r},t) = m\, n(\boldsymbol{r},t)$ and $\rho_q(\boldsymbol{r},t) = q\, n(\boldsymbol{r},t)$
Perpendicular and parallel pressures (refer to relations (4.17)): $p_\perp = 1/2\, mn\, (\langle v_\perp^2 \rangle - \langle v_\perp \rangle^2)$ and $p_\parallel = mn(\langle v_\parallel^2 \rangle - \langle v_\parallel \rangle^2)$
Magnetic moment density (refer to (4.24)): $\boldsymbol{M} = -1/2mn\langle (\boldsymbol{v}_\perp - \boldsymbol{V}_D)^2 \rangle\, \boldsymbol{B}/B^2 = -p_\perp/B\, \boldsymbol{e}$
Angular momentum density: $\boldsymbol{L} = (2m/q)\, \boldsymbol{M}$

For the bulk velocity things are different. Each virtual guiding center particle has a perpendicular drift velocity $\boldsymbol{V}_D$ which, however, is not an independent variable: it is a function of $v_\perp$, $v_\parallel$ and the local magnetic field (2.14). On the other hand, the parallel velocity of a guiding center particle is a vector equal to the original particle's parallel velocity $\boldsymbol{v}_\parallel$, and it is an independent variable (2.11). Thus for a guiding center fluid we write:

Bulk perpendicular (or drift) velocity: $\boldsymbol{V}_{g\perp} = \int \boldsymbol{V}_D F\, dv_\perp dv_\parallel / \int F\, dv_\perp dv_\parallel$[3]
Bulk parallel (or field-aligned) velocity: $\boldsymbol{V}_{g\parallel} = \int \boldsymbol{v}_\parallel F\, dv_\perp dv_\parallel / \int F\, dv_\perp dv_\parallel$

As mentioned above, we will mainly deal with ensembles with symmetric pitch angle distributions in which there is no field-aligned bulk streaming (no field-aligned convection currents) and $V_{g\parallel} = \langle \boldsymbol{v}_\parallel \rangle \equiv 0$.

It is important to understand the difference between the bulk velocities in both fluid models. In the kinetic model, $\boldsymbol{V}$ is the spatial average of the instantaneous velocity vectors of *actual* particles in an element of volume, whereas in the GC fluid, $\boldsymbol{V}_g$ is a double average: the spatial average of the velocities $\boldsymbol{V}_D$, $\boldsymbol{V}_\parallel$, which are averages (over a cyclotron turn) of the velocity components of a particle: $\boldsymbol{V}_g = \langle \langle \boldsymbol{v} \rangle_\varphi \rangle$. The "missing part" of particle motion in the GC fluid model is encoded in the magnetic moment of each GC particle. The bulk velocity vector $\boldsymbol{V}_g$ of an ensemble of guiding center particles always describes *true* macroscopic mass transport, whereas the mean velocity vector $\boldsymbol{V}$ of the ensemble of the original particles may not—both velocity vectors in general will differ from each other

[3]From now on, all macro-variables in the guiding center fluid will carry the subindex g, whereas homologous variables in the kinetic particle fluid model will not be subindexed.

($\boldsymbol{V}_g \neq \boldsymbol{V}$); in fact, they even may be opposite to each other. Below we'll show some simple examples.

What both models *must* have in common, are the values of local current density $\boldsymbol{J}$, which links the ensemble dynamically with the local magnetic field (action of the Lorentz force $\boldsymbol{J} \times \boldsymbol{B}$ *on* the current, and contribution of the current *to* the sources $\nabla \times \boldsymbol{B}$ of the magnetic field). In the kinetic fluid model, which doesn't care whether or not individual particles in a given element of volume have mesoscopically organized motion (such as cyclotron gyration), the current density is $\boldsymbol{J} = \rho_q \boldsymbol{V}$, a pure *convection* current. In the guiding center fluid, we must use the usual $E\&M$ expression for magnetized media taking into account the *equivalent* currents $\boldsymbol{J}_{eq} = \nabla\times\boldsymbol{M}$ (Appendix A.1). The total electric current density will thus be:

$$\boldsymbol{J} = \rho_q \boldsymbol{V}_g + \nabla\times\boldsymbol{M} \tag{5.3}$$

and consequently[4]

$$\boldsymbol{V} = \boldsymbol{V}_g + {\rho_q}^{-1}\nabla\times\boldsymbol{M} = \boldsymbol{V}_g - {\rho_q}^{-1}\nabla\times\left(\frac{p_\perp}{B^2}\boldsymbol{B}\right) = \boldsymbol{V}_g + 1/2{\rho_m}^{-1}\nabla\times\boldsymbol{L} \tag{5.4}$$

The last equality stems from the definition of angular momentum density

$$\boldsymbol{L} = 2(m/q)\,\boldsymbol{M} \tag{5.5}$$

We end this section with the promised discussion of some "kindergarten" examples, to show in semi-quantitative form that the perpendicular component $\boldsymbol{V}_\perp$ of the bulk velocity of a kinetic fluid is indeed not necessarily equal to the perpendicular bulk velocity of the corresponding guiding center fluid. Quite generally, these examples are intended to shed some light on the physical nature of different, distinct classes of currents in a guiding center fluid. Consider Fig. 5.1 left side, which depicts a gyrotropic ensemble of mono-energetic 90° pitch angle particles, with a particle density gradient ∇n in the direction of the x-axis, cycling in a uniform external magnetic field $\boldsymbol{B}$ directed along the z-axis. If we do a *cyclotron* average of the perpendicular velocity vector of any given particle to obtain its guiding center drift velocity (1.3), we obviously get $\boldsymbol{V}_D = \langle \boldsymbol{v}_\perp \rangle_{cyclotron} = 0$. The particles are all gyrating in situ and the guiding centers are all at rest—there is no flow in the guiding center fluid and there is no net transport of mass or electric charge: the guiding center convection current is $\boldsymbol{J}_g = \rho_q \boldsymbol{V}_g = 0$. But in the guiding center fluid there also will be an *equivalent* current $\boldsymbol{J}_{eq} = \nabla\times\boldsymbol{M} = \nabla\times(-1/2mnv_\perp{}^2/B)\,\boldsymbol{e} = -(1/2mv_\perp{}^2/B)\,\nabla n\times\boldsymbol{e} \neq 0$, always in the direction of $+y$ (regardless of the particles' charge q). This (and the next set) describes the effect

[4]The following relation (5.4) can be deduced directly for gyrotropic ensembles by linking the distribution functions f (4.5) and F (5.2) using (1.25) and the definitions of $\boldsymbol{V}$, $\boldsymbol{V}_g$ and $\boldsymbol{M}$ (the proof is lengthy!).

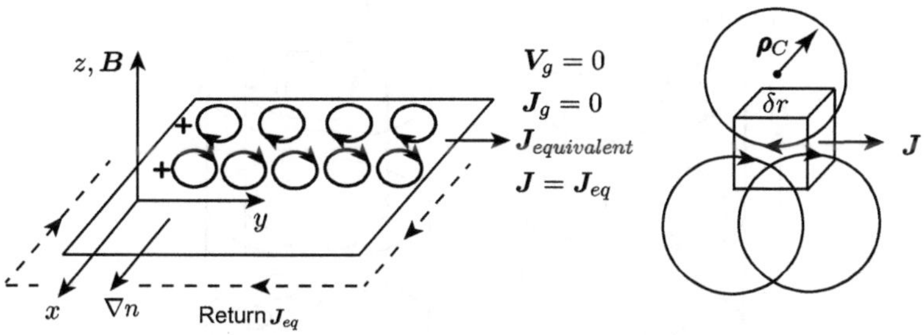

Fig. 5.1 Sketch showing the physical origin of an equivalent current density in a non-uniform distribution of 90° pitch angle particles (zero GC current, no net charge transport)

of the magnetic field on a specific distribution of guiding center particles; later we will discuss the effect of specific ensembles (their currents) on the magnetic field.

We now turn to the kinetic model description of the same ensemble of Fig. 5.1. For this, we must look at the figure "with a magnifying glass" (or, more realistically, with a very small detector) and realize that the individual particle distribution will be anisotropic: considering a domain much smaller than ${\rho_C}^3$ (right side of the figure) we will always see (or detect) more particles traveling in the $+y$ direction than in any other. This represents a local *convection* current $\boldsymbol{J}$ along the y-axis.[5] Since we demand that both fluid model descriptions must be consistent with each other in terms of their macroscopic electromagnetic manifestations, for the case of Fig. 5.1 the convection current in the kinetic fluid must be the same as the equivalent current in the guiding center fluid: $\boldsymbol{J} = \nabla \times \boldsymbol{M}$.

In the example of Fig. 5.1 there is no net transport of electric charge ($\rho_q = 0$, $\partial \rho_q / dt = 0$), yet there is a current density everywhere inside the ensemble. Obviously, conservation of charge tells us that $\nabla \cdot \boldsymbol{J} = 0$, so both, the equivalent currents and (in the particle fluid picture) the convection currents must be *closed* somewhere. Observe Fig. 5.1 (left side): the distribution of particles does not extend to infinity—it must have a boundary somewhere along the x and y axes, which means that, eventually, somewhere there must be negative number density gradients. Such gradients represent current densities, precisely the ones that close the J_y current system in the above example, as sketched in the figure. This observation shows that quite generally it is extremely dangerous to speculate qualitatively about current systems in the magnetosphere (e.g., about the neutral sheet current) without explicitly including a precise picture of all closing currents, too (e.g., the current system where the neutral sheet merges into the tail boundary).

[5]Remember that this is the usual explanation given in $E\&M$ texts to justify the appearance of an equivalent $\boldsymbol{\nabla} \times \boldsymbol{M}$ current in magnetized materials (although in ferromagnetism the magnetization is not due to "little current loops" in atoms but due to the intrinsic quantum magnetic moment (spin) of electrons). Since in an ensemble of trapped particles the magnetic moment associated to a guiding center particle is always directed antiparallel to B, plasmas behave like a diamagnetic gas—as we already had anticipated.

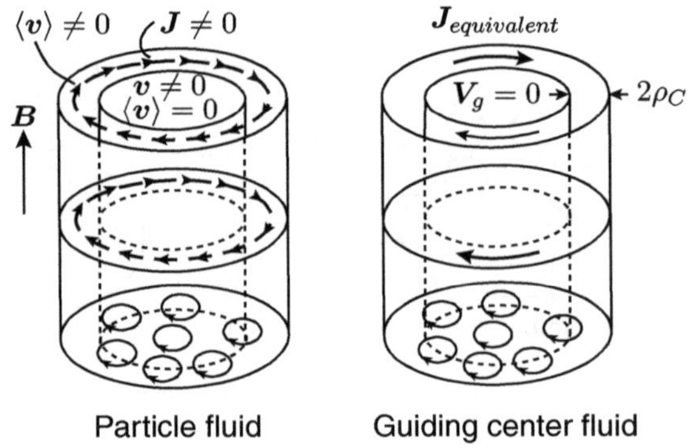

Fig. 5.2 Flux cylinder in a uniform field with a uniform distribution of positive, mono-energetic 90° pitch angle particles. *Left*: Viewed as a guiding center fluid; there are no flows anywhere, but an equivalent current $\nabla \times M$ in the boundary layer due to magnetic moment cut-off. *Right*: Viewed as a particle fluid at the microscopic level; laminar flow within $2\rho_C$ of the outer boundary

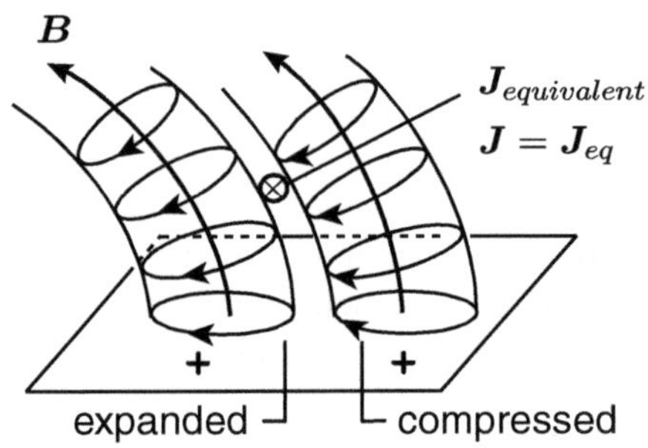

Fig. 5.3 Origin of the "bent sausage" equivalent current

To consolidate understanding of the case of Fig. 5.1, consider a cylinder of field lines in a uniform magnetic field (a magnetic flux tube) filled with 90° mono-energetic particles with uniform guiding center density, Fig. 5.2. Viewed as a particle fluid, the distribution inside the cylinder will be isotropic everywhere, with zero average velocity except in a thin boundary layer of thickness $\delta r = 2\rho_C$, where there will be a laminar flow (macroscopically a surface current). Viewed as a guiding center particle fluid, the velocity inside the cylinder will be zero, too, but now it is the sudden jump to zero of the magnetization density in the boundary layer (due to the cut-off of guiding center density) which will lead to an *equivalent* current that must be equal to the surface convection current in the kinetic model description. Note that in this case, the current system is closed in itself.

Next consider the example of 90° pitch angle particles uniformly distributed along curved field lines as shown in Fig. 5.3, with B nearly constant along and across those field lines. Such a situation is, indeed, highly artificial and can only represent an instant snapshot: the mirror force (2.8) would immediately start moving the particles along $\boldsymbol{B}$—anyway, this is just a kindergarten example! Viewed either as a guiding center fluid or a particle fluid, the respective average velocities $V_{g\|}$

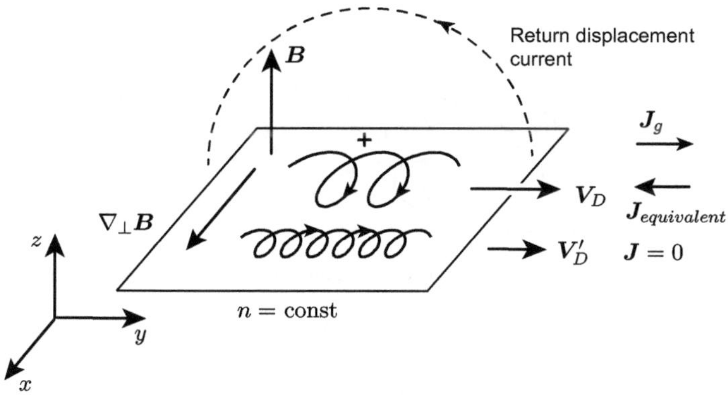

Fig. 5.4 Origin of the equivalent counter-current J_{eq} in a uniform particle distribution of gradient-B drifting particles

and $V_\parallel$ along the field lines are assumed to be zero (no field-aligned current). But there should be an equivalent current perpendicular to $\boldsymbol{B}$ (into the plane of the figure), of value $\boldsymbol{J}_{eq} = \nabla\times(-p_\perp/B)\boldsymbol{e} = p_\perp/B\,(\partial\boldsymbol{e}/\partial s \times \boldsymbol{e})$ (remember that $\partial\boldsymbol{e}/\partial s = -\boldsymbol{n}/R_C$, with R_C the field line radius of curvature, relation (A.15) in Appendix A.1). Its origin is simple: just look at the figure with a magnifying glass, and you'll see that positive particles move toward you in their cyclotron motion on the convex side of the field line, and go into the paper in a slightly *compressed* fashion, on the concave side. In other words, a tiny detector would see, per unit surface, more particles going into the paper than coming out of it. This is yet another case in which $\boldsymbol{J} = \nabla\times\boldsymbol{M}$ (perpendicular components only!).

Another example, sketched in Fig. 5.4, is that of a uniform 90° pitch angle mono-energetic particle distribution on the minimum-B surface of a magnetic field with a constant gradient $\nabla_\perp B \neq 0$ in the x direction, but *no* gradient in number density n. Guiding centers will drift with velocity $\boldsymbol{V}_g$ to the right along the x-axis, which represents a guiding center convection current to the right $\boldsymbol{J}_g = -p_\perp/B^2\,\nabla_\perp B\times\boldsymbol{e}$. In addition, there will be an equivalent current $\nabla\times\boldsymbol{M}$, where $\boldsymbol{M} = -(p_\perp/B)\,\boldsymbol{e}$. Since for a uniform distribution only $\boldsymbol{B}$ depends on the position $\boldsymbol{r}$ this equivalent current is directed to the left and *exactly cancels* the convection drift current $\boldsymbol{J}_g$ (remember that $\nabla \times \boldsymbol{e} = -1/B(\nabla B \times \boldsymbol{e})$, so that in this case the total current density in the guiding center fluid model is zero. Therefore, the convection current in the particle fluid model should also be zero. It is a little trickier to convince oneself, by looking at the figure, that the pertinent velocity distribution of particles on the minimum-B surface is indeed isotropic, and that the number of particles moving in the $+y$ direction in a very small element of volume is always the same as the number of those going in the opposite direction, for constant $p_\perp$. In summary, in the case of Fig. 5.4, the current density in the particle fluid is $\boldsymbol{J} = \boldsymbol{J}_g + \nabla\times\boldsymbol{M} \equiv 0$. Here we have an example of a particle distribution with net mass and charge transport (to the right in the figure), but in which the local average particle velocity is zero (the implication of this fact for the magnetospheric ring current will be discussed

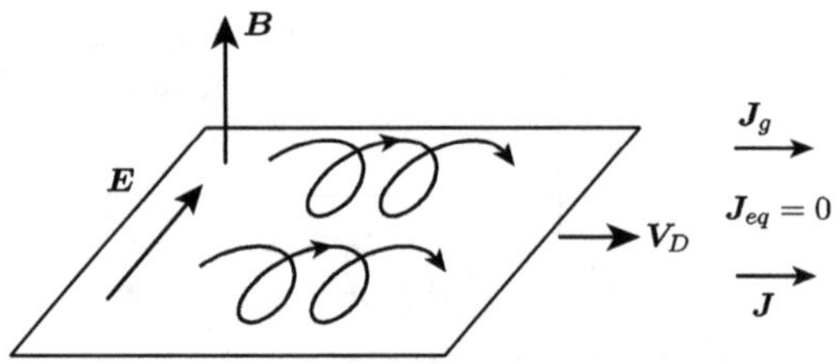

Fig. 5.5 Case of a GC current equal to the convection current (zero equivalent current)

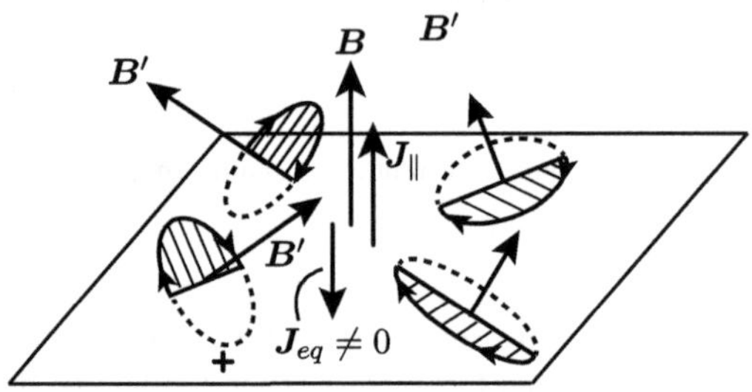

Fig. 5.6 Sketch (not in scale!) of positive particles on field lines twisted by a parallel current along the central axis (snapshot from the guiding center system moving with velocity $V_{\|}$). The cyclotron loops are perpendicular to the local, inclined, $\boldsymbol{B}'$. Near the central axis there is a net uncompensated velocity component from the cyclotron motion of the particles into the plane $\perp$ to $\boldsymbol{B}$

briefly in the next section). Note that in this example there is no equivalent current to close because there just is no current. It is an interesting exercise to figure out the electrodynamics near the right edge of the drifting particle distribution (something quite relevant to the physics of magnetospheric plasma blobs!)

A uniform gyrotropic distribution of mono-energetic particles in a uniform magnetic field, crossed by a uniform electric field $\boldsymbol{E}$ (Fig. 5.5), is a typical example where $\nabla\times\boldsymbol{M} \equiv 0$, i.e., in which the convection currents in both models are equal: $\boldsymbol{J} = \boldsymbol{J}_g$. Note however the difference in physical character of the two types of currents: the particle current $\boldsymbol{J}$ in the kinetic model is due to the fact that, because of electric field acceleration, particles moving to the right in the figure are on the "upper" arc of the cyclotron orbit where their speed is higher than that of left-moving lower arc particles (see (1.35) and Fig. 1.7). Guiding center particles drift at constant speed, equal to the average speed of the actual particles in their OFR.

Our last kindergarten example is that of a configuration in which the mono-energetic particles have a pitch angle $<90°$ in a near-uniform magnetic field. Now there will be a component of $\boldsymbol{J}_g$ parallel to $\boldsymbol{B}$. As explained in Appendix A.1, any parallel current causes a twist of the magnetic field lines: $\nabla\times\boldsymbol{M}$ will have a component parallel to $\boldsymbol{B}$. Consider Fig. 5.6, shown in the guiding center frame of reference that moves with the velocity $\boldsymbol{v}_{\|}$ (assumed common to all particles in this example) along the central field line. As sketched in the figure, because of the torsion, neighboring cyclotron loops are slanted with respect to those of particles circling the central field line. As a result, a geometry arises such that near the axis of symmetry the neighboring particles have a component of motion *counter* the main parallel drift whenever they cross the central loops, a motion that is uncompensated by upward-moving particles from neighboring cyclotron loops. In other words, a parallel equivalent current $\nabla\times\boldsymbol{M}|_{\parallel}$ will always counteract, i.e., diminish the effect of whatever field-aligned guiding center current flows in the first place.

The discussion of these five simple examples suggests that it is always important *to clearly state which fluid model*, kinetic or guiding center, is being invoked in a plasma[6] description. In theoretical analysis and numerical calculations one mostly works with the kinetic model, but when describing intuitively and qualitatively a plasma system, or even when trying to interpret measurement results, physicists more often than not are thinking about, or visualizing, the system in guiding center model terms (often without explicitly saying so). However, there is a danger of using the guiding center fluid model, despite its greater physical intuitiveness concerning the ensemble's macroscopic properties such as electric currents and internal stresses (see next section). One too often forgets that this model is valid only provided that: (i) the guiding center approximations (2.1) and (2.2) apply everywhere (thus excluding neutral lines, sharp gradients, rapid oscillations, etc.), (ii) the collision rate is negligible (thus excluding the ionospheric regions), and (iii) particle distributions are gyrotropic (thus excluding waves in the cyclotron frequency or higher range).

5.3 Currents and Stresses Arising from Interactions with the Magnetic Field

In order to analyze the types of currents sustained by an ensemble of guiding center particles defined by a distribution function $F(\boldsymbol{r}, t, v_\perp, v_\parallel)$, we turn to the general expression of the drift velocity (2.14) given in Sect. 2.1. We shall assume that the guiding center approximation is valid everywhere and that there is no bounce bunching ($\langle \boldsymbol{v}_\parallel \rangle = 0$), and we shall neglect the action of external non-electromagnetic forces ($\boldsymbol{F} \equiv 0$) as well as all higher order drift terms, i.e., we shall retain only the 1st, 3rd and 4th terms. We thus write, for the transverse drift velocity vector of a single guiding center particle:

$$\boldsymbol{V}_{g\perp} = \boldsymbol{V}_D = \left[q\boldsymbol{E} - \frac{mv_\perp^{*2}}{2B}\nabla B - mv_\parallel^2\frac{\partial \boldsymbol{e}}{\partial s} - m\frac{d\boldsymbol{V}_D}{dt}\right] \times \frac{\boldsymbol{B}}{qB^2}$$

Remember that $v_\perp^*$ is the modulus of the perpendicular component of the actual particle's velocity in its guiding center system at the point in question ($\boldsymbol{v}_\perp{}^* = \boldsymbol{v}_\perp - \boldsymbol{V}_D$, relation (1.6)). It is evident that in the bracket, only the perpendicular

[6] Never mind that in these kindergarten examples we have considered only one class of particles—the results about currents thus far are independent of the electric charge of the particles involved.

components of $\boldsymbol{E}$ and $d\boldsymbol{V}_D/dt$ will contribute. The parallel guiding center velocity $\boldsymbol{V}_{g\|}$ will be equal to the cyclotron-average of the parallel velocity of the actual particle (2.11), but as mentioned above, for the time being we will exclude streaming along field lines ($\boldsymbol{V}_{g\|} \equiv 0$).

We now use this equation to determine the macroscopic quantity $\langle \boldsymbol{V}_D \rangle$ (transverse bulk velocity), by multiplying each term by the guiding center distribution function $F(\boldsymbol{r}, t, v_\perp, v_\|)$ and integrating over velocity space $[v_\perp, v_\|]$. We obtain, taking into account the definitions (4.17) of $p_\perp$ and $p_\|$ (the latter, with $\boldsymbol{V}_{g\|} = 0$):

$$\boldsymbol{V}_{g\perp} = \langle \boldsymbol{V}_D \rangle = \left[\rho_q \boldsymbol{E} - \frac{p_\perp}{B} \nabla_\perp B - p_\| \frac{\partial \boldsymbol{e}}{\partial s} - \rho_m \frac{d\boldsymbol{V}_{g\perp}}{dt} \right] \times \frac{\boldsymbol{B}}{\rho_q B^2} \tag{5.6}$$

To obtain the total current $\boldsymbol{J}$, we have to multiply this equation with ρ_q and add $\nabla \times \boldsymbol{M}$. This latter magnetization current density is

$$\nabla \times \boldsymbol{M} = \nabla \times \left(-\frac{p_\perp}{B^2} \boldsymbol{B} \right) = -\frac{p_\perp}{B^2} \nabla \times \boldsymbol{B} - \nabla p_\perp \times \frac{\boldsymbol{B}}{B^2} + 2 \frac{p_\perp}{B} \nabla B \times \frac{\boldsymbol{B}}{B^2}$$

The second and third terms are perpendicular to $\boldsymbol{B}$. Using (A.26) and (A.27) of Appendix A.1 for $\nabla \times \boldsymbol{B}|_\perp$ and $\nabla \times \boldsymbol{e}|_\perp = \boldsymbol{e} \times \partial \boldsymbol{e}/\partial s$, we obtain for the $\perp$ and $\|$ components of the equivalent current density:

$$\nabla \times \boldsymbol{M}|_\perp = \left[-\nabla_\perp p_\perp + p_\perp \frac{\partial \boldsymbol{e}}{\partial s} + p_\perp \frac{\nabla_\perp B}{B} \right] \times \frac{\boldsymbol{B}}{B^2} \tag{5.7}$$

$$\nabla \times \boldsymbol{M}|_\| = -\frac{p_\perp}{B^2} \nabla \times \boldsymbol{B}|_\| \tag{5.8}$$

Turning first to the transverse equations, we add expressions (5.7) and (5.6) (multiplied by ρ_q) to obtain for the total transverse current density $\boldsymbol{J}_\perp$ (which of course must also be the total transverse current density in the corresponding particle fluid model):

$$\begin{aligned} \boldsymbol{J}_\perp &= \boldsymbol{J}_{g\perp} + \nabla \times \boldsymbol{M}|_\perp \\ &= \left[\rho_q \boldsymbol{E} - \nabla_\perp p_\perp - (p_\| - p_\perp) \frac{\partial \boldsymbol{e}}{\partial s} - \rho_m \frac{d\boldsymbol{V}_{g\perp}}{dt} \right] \times \frac{\boldsymbol{B}}{B^2} \\ &= \boldsymbol{J}_E + \boldsymbol{J}_D + \boldsymbol{J}_A + \boldsymbol{J}_I \end{aligned} \tag{5.9}$$

where

$$\boldsymbol{J}_E = \rho_q \frac{\boldsymbol{E}}{B} \times \boldsymbol{e} \qquad \text{Electric field drift current} \tag{5.10}$$

$$\boldsymbol{J}_D = -\frac{\nabla_\perp p_\perp}{B} \times \boldsymbol{e} \qquad \text{diamagnetic current} \tag{5.11}$$

$$\boldsymbol{J}_A = -\frac{(p_\parallel - p_\perp)}{B}\frac{\partial \boldsymbol{e}}{\partial s} \times \boldsymbol{e} \qquad \text{“pressure anisotropy” current} \tag{5.12}$$

$$\boldsymbol{J}_I = \frac{\rho_m}{B}\frac{d\boldsymbol{V}_{g\perp}}{dt} \times \boldsymbol{e} \qquad \text{Inertial current} \tag{5.13}$$

Notice that an important rearrangement has taken place by adding the equivalent current that arises from the magnetization of the guiding center fluid. In particular, the gradient-B drift current term (second one in (5.6)) *has dropped out*, cancelled by an homologous term in (5.7). This is exactly what happened in our "kindergarten example" shown in Fig. 5.4 in the preceding section![7] In the above list, $\boldsymbol{J}_E$ and $\boldsymbol{J}_I$ are convection currents, $\boldsymbol{J}_D$ is an equivalent current and $\boldsymbol{J}_A$ is mixed: the first part (with $p_\parallel$) is a convection current carried by the field-line curvature drift (see (2.14)) whereas the second part (with $p_\perp$) is the equivalent current whose microscopic origin was shown in the "kindergarten" example of Fig. 5.3. Notice that for isotropic pressure ($p_\parallel = p_\perp$) $\boldsymbol{J}_A \equiv 0$, which means that also the curvature drift drops out, cancelled by the (unnamed) equivalent current part of $\boldsymbol{J}_A$. All currents depend on $\boldsymbol{B}$ and the particle distribution (pressure tensor or density): the local magnetic field dictates, and the particle ensemble properties drive, the currents! Note that in (5.6) only the electric field drift is independent of the particles' properties; thus it will not contribute to the total current density in a collisionless *charge-neutral* ensemble of two or more species.

An important point is that relations (5.9)–(5.12) are valid in both fluid models, the kinetic and the guiding center one. Concerning relation (5.13), it can be shown (rather laboriously), that although in general $\boldsymbol{V}_g \neq \boldsymbol{V}$, for the total time derivatives $d\boldsymbol{V}_g/dt \cong d\boldsymbol{V}/dt$ within the guiding center approximation, so that this relation (5.13) is valid, too, in both models. The current (5.9) is thus indeed the total current density that acts as the *source* of a magnetic field, i.e., the one that enters in Maxwell's equations (A.49). This somewhat trivial remark will be important later.

Concerning the parallel bulk velocity, let us lift for a moment the initial assumption that it is zero. We shall have

$$\boldsymbol{V}_{g\parallel} = \langle \boldsymbol{v}_\parallel \rangle, \tag{5.14}$$

basically an independent variable in the sense that at one given point it *only* depends on the particle distribution function there—which, however, as we will show in the next section, varies in a specific manner along any given magnetic field line. For an equation for $\boldsymbol{J}_\parallel$, complement to (5.9), we write

[7] This dropout, predicted by theoreticians in the early days of magnetospheric physics, caused confusion among experimentalists studying ring current data, who from the beginning assumed this West-East current to be due to the *convective* E-W and W-E drift of trapped protons and electrons, respectively. However, the ring current is the superposition of a E-W convection drift current with an equivalent *diamagnetic* current (5.11), the latter with an W-E inner ring (radially outward directed density or pressure gradient in (4.24)) and a E-W outer ring where the density gradient is reversed.

$$\boldsymbol{J}_{\parallel} = \rho_q \boldsymbol{V}_{g\parallel} + \nabla \times \boldsymbol{M}|_{\parallel} = \rho_q \boldsymbol{V}_{g\parallel} - p_{\perp}/B^2 \nabla \times \boldsymbol{B}|_{\parallel}$$

or, taking into account that under stationary or slowly varying conditions $\nabla \times \boldsymbol{B}|_{\parallel} = \mu_0 \boldsymbol{J}_{\parallel}$, where $\boldsymbol{J}_{\parallel}$ is the total field-aligned current,

$$\boldsymbol{J}_{\parallel} = \boldsymbol{J}_{g\parallel} \frac{1}{1 + p_{\perp}/(B^2/\mu_0)} \tag{5.15}$$

Note that always $J_{\parallel} \leq J_{g\parallel}$. This confirms what we have anticipated in the fifth kindergarten example of the previous section (see Fig. 5.6). The equation shows that if $V_{g\parallel} = 0$ (no average parallel velocity of guiding centers, symmetric pitch angle distribution), the total field-aligned current density is always zero—in other words, a field-aligned current cannot "be made of" an equivalent current alone. If on the other hand, there is GC field-aligned streaming ($\boldsymbol{J}_{g\parallel} \neq 0$) and the transverse particle pressure $p_{\perp}$ is much smaller than the magnetic energy density $B^2/2\mu_0$ (Appendix A.1, relation (A.40)—the ratio p/u is called the *beta* of the plasma), which in general implies low particle number density, the total field-aligned current density $J_{\parallel}$ is maximum and equal to it. If in the other extreme $p_{\perp} \sim B^2/2\mu_0$ (high number density), $J_{\parallel} \to 0$ again, regardless of the parallel streaming of guiding centers (cancelled by the equivalent current in the guiding center model). This is an example of the above-mentioned special nature of parallel motions in the fluid descriptions. As we have seen in Appendix A.1, the field-aligned current is responsible for a twist of magnetic field lines; in the present example it also controls the proportion between convection and equivalent currents in the guiding center fluid. But remember that field-aligned currents, while causing torsion in the magnetic field ((A.28) and (A.29)), cannot sustain any magnetic stresses: $\boldsymbol{J}_{\parallel} \times \boldsymbol{B} \equiv 0$.

We now turn to the general stresses, i.e., the average macroscopic Lorentz force densities acting inside the guiding center fluid, $\boldsymbol{J} \times \boldsymbol{B} = \boldsymbol{J}_{\perp} \times \boldsymbol{B} = (\boldsymbol{J}_E + \boldsymbol{J}_D + \boldsymbol{J}_A + \boldsymbol{J}_I) \times \boldsymbol{B}$ (5.9). Let us begin again with the kindergarten example of 90° pitch angle particles filling a cylindrical flux tube in a uniform $\boldsymbol{B}$-field (Fig. 5.2). Regardless of the fluid model considered (left or right in the figure), there will be a thin layer of current on the surface of the cylinder, as shown in that figure. If there are "many, many" particles, two things will happen: (i) the magnetic field inside the cylinder will decrease noticeably due to the solenoidal surface currents (diamagnetic property of the ensemble), and (ii) an average non-negligible Lorentz force will appear acting on the outer equivalent current-carrying part of the ensemble. This latter outward-directed force density $\boldsymbol{J} \times \boldsymbol{B}$ represents an internal *stress* in the ensemble, quite similar to the magnetostriction acting on equivalent $\nabla \times \boldsymbol{M}$ currents inside condensed matter with magnetization density $\boldsymbol{M}$. Our kindergarten example can be carried further qualitatively: as the magnetic field in the cylinder decreases with time, an induced electric field will appear (see example with the case of an increasing field on page 21!) and the associated outward drift will expand the particle ensemble. But equivalently, in the GC model we could well attribute the expansion to the action of an outward Lorentz force stress! A similar quantitative analysis can be made with the example of Fig. 5.6: here we have a convection current density; the average Lorentz force on it counteracts exactly the electric field force—a good example to convince a skeptic that a plasma exposed to an electric field is *not* accelerated in the direction of the field but will drift perpendicularly to it!

If we cross Eq. (5.9) with $\boldsymbol{B}$ and rearrange terms, we are led to the following *dynamic equation for the perpendicular bulk flow* in a guiding center fluid:

$$\rho_m \left. \frac{d\boldsymbol{V}_g}{dt} \right|_{\perp} = \rho_q \boldsymbol{E}_\perp - \nabla_\perp p_\perp - (p_\parallel - p_\perp)\frac{\partial \boldsymbol{e}}{\partial s} + \boldsymbol{J}_\perp \times \boldsymbol{B} \tag{5.16}$$

Observe that this is not a "true" dynamic equation which, by integration, would lead to the calculation of $\boldsymbol{V}_g$; it merely serves to display the stresses or force densities responsible for the transverse acceleration of parcels in the guiding center fluid model.

To derive a dynamic equation of flow parallel to the magnetic field, complement to Eq. (5.16), we can convert the single-particle parallel equation (2.20) (without non-electric forces) into a macroscopic equation for an ensemble of particles by multiplying it with the guiding center distribution function F and integrating, to obtain:

$$\rho_m \left. \frac{d\boldsymbol{V}_g}{dt} \right|_{\parallel} = \rho_m \frac{d\boldsymbol{V}_g}{dt} \cdot \boldsymbol{e} = \rho_q \boldsymbol{E}_\parallel - \frac{\partial p_\parallel}{\partial s} + \frac{(p_\parallel - p_\perp)}{B}\frac{\partial B}{\partial s} \tag{5.17}$$

The two above equations can be combined into one by taking into account (4.21):

$$\rho_m \frac{d\boldsymbol{V}_g}{dt} = \rho_q \boldsymbol{E} - \nabla \mathbb{P} + \boldsymbol{J} \times \boldsymbol{B} \tag{5.18}$$

This general equation for the bulk velocity of an ensemble of guiding centers explicitly reveals the dynamic action of three types of physical causes: (i) non-magnetic forces (the first term, to which any non-electromagnetic force density could be added), (ii) the "mechanical stresses" represented by the pressure tensor, and (iii) the action of the magnetic field on the ensemble through Lorentz-type forces (the third term). It is important to remember that at this stage of our discussion, we are still dealing with just *one species* of particles and that the magnetic and electric fields are given, i.e., that the contribution to the fields of charges and currents in the ensemble are being neglected.

Before we get real and drop this limitation, we end this section with an analysis of the physical meaning of the stresses in a guiding center fluid, as illustrated by the hypothetical example of a "magnetohydrostatic" equilibrium state of the ensemble: no time dependence, no total current density, no external forces, no electric field. The condition of equilibrium implies that transverse equation (5.16) now should be:

$$\nabla_\perp p_\perp + (p_\parallel - p_\perp)\frac{\partial \boldsymbol{e}}{\partial s} = 0$$

This equation is identical to (4.33), only that here it has been derived in a more general way. According to (5.9), this equation also implies that $\boldsymbol{J}_D + \boldsymbol{J}_A = 0$ in the case of a stationary state.

Along the respectively binormal and normal x and y axes (in the natural coordinate system, Appendix A.1),

$$\frac{\partial p_\perp}{\partial x} = 0$$

$$\frac{\partial p_\perp}{\partial y} - (p_\parallel - p_\perp)\frac{1}{R_c} = 0$$

where $R_c = |\partial \boldsymbol{e}/\partial s|^{-1}$ is the field line's radius of curvature. These two equations show how the perpendicular pressure and the pressure anisotropy $p_\parallel - p_\perp$ must obey stringent conditions of spatial variability in a magnetostatic field to remain in equilibrium (indeed, it is useful to re-examine the example given in Sect. 4.4, where the equilibrium conditions were derived for a specific case.) Under the same static equilibrium conditions, the following relation is obtained for the parallel stresses from (5.17), in partnership with the transverse equation (4.33):

$$\frac{\partial p_\parallel}{\partial s} - \frac{(p_\parallel - p_\perp)}{B}\frac{\partial B}{\partial s} = 0$$

This equation is identical to (4.32).

To interpret the detailed physical meaning of the various terms in the above equilibrium relations (4.33) and (4.32), consider a guiding center fluid element in a magnetic flux tube, as sketched in Fig. 5.7. In this figure, the axes represent the natural reference frame, with $z \parallel \boldsymbol{e}$, y along the normal $\boldsymbol{n}$ and x along the binormal $\boldsymbol{b}$ (Appendix A.1). As always, R_C is the radius of curvature of the field lines, and we have the following relations between the side areas: $\delta A_x^* = \delta A_x$ (binormal axis); $\delta A_y^* = \delta A_y(1 + \delta y/R_C)$ (field-geometric factor) and $\delta A_z^* = \delta A_z(1 - 1/B\,(\partial B/\partial s)\,\delta s)$ (conservation of magnetic flux).

Along the y-axis, the following forces act on the GC fluid element in hydrostatic balance:

$$-(p_\perp + \frac{\partial p_\perp}{\partial y}\delta y)\delta A_y(1 + \frac{\delta y}{R_c}) + p_\perp \delta A_y + p_\parallel\left|\frac{\partial \boldsymbol{e}}{\partial s}\right|\delta A_y \delta y = 0$$

in which the third term represents the total centrifugal force on the particles. This means that

$$\frac{\partial p_\perp}{\partial y} - (p_\parallel - p_\perp)\left|\frac{\partial \boldsymbol{e}}{\partial s}\right| = 0$$

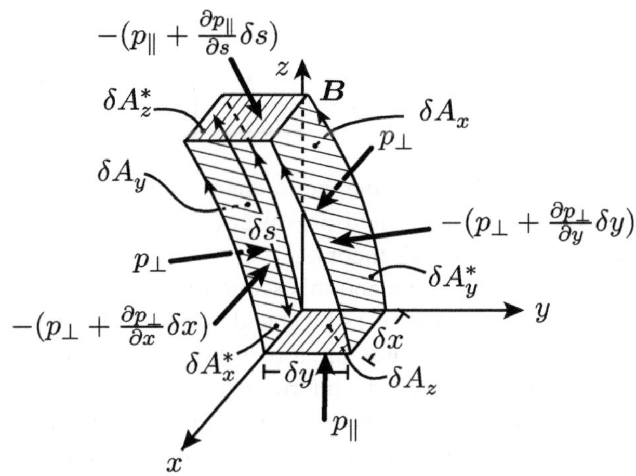

Fig. 5.7 Flux tube element filled with guiding center fluid in "hydrostatic" equilibrium

To find the other term $\partial p_\perp/\partial x$, we have along the x axis:

$$-\left(p_\perp + \frac{\partial p_\perp}{\partial x}dx\right)\delta A_x + p_\perp \delta A_x = 0$$

which means that

$$\frac{\partial p_\perp}{\partial x} = 0$$

The above relations represent the vector equilibrium condition (4.33) along the y and x axes perpendicular to $\boldsymbol{B}$.

For the parallel equation of equilibrium (along the z-axis), consider again Fig. 5.7. With the total mirror force density on the guiding center particles in the flux tube element $(-p_\perp/B\ (\partial B/\partial s))$, we have the following equilibrium condition:

$$-\left(p_\parallel + \frac{\partial p_\parallel}{\partial s}\delta s\right)\left(1 - \frac{1}{B}\frac{\partial B}{\partial s}\delta s\right) + p_\parallel - \frac{p_\perp}{B}\frac{\partial B}{\partial s}\delta s = \left(-\frac{\partial p_\parallel}{\partial s} + \frac{(p_\parallel - p_\perp)}{B}\frac{\partial B}{\partial s}\right)\delta s = 0$$

leading to (4.32).

Finally, we may relax a bit the a priori conditions in our example and admit transverse currents that, however, are independent of time—i.e., limited to a stationary ensemble. This does not change the parallel equation (4.32) and its physical meaning (Fig. 5.7). The perpendicular equation (5.16) now becomes, taking into account (A.36) of Appendix A.1:

$$\nabla_\perp p_\perp + (p_\parallel - p_\perp)\frac{\partial \boldsymbol{e}}{\partial s} = \boldsymbol{J}_\perp \times \boldsymbol{B} = \frac{B^2}{\mu_0}\frac{\partial \boldsymbol{e}}{\partial s} - \nabla_\perp\left(\frac{B^2}{2\mu_0}\right) \tag{5.19}$$

or:

$$\nabla_\perp\left(p_\perp + \frac{B^2}{2\mu_0}\right) + \left(p_\parallel - p_\perp - \frac{B^2}{\mu_0}\right)\frac{\partial \boldsymbol{e}}{\partial s} = 0 \tag{5.20}$$

In this equation, the magnetic energy density $B^2/2\mu_0$ (A.40) plays the role of transverse magnetic field pressure; $(B^2/\mu_0)\ \partial \boldsymbol{e}/\partial s$ is the perpendicular magnetic tension of Maxwell's theory.

To summarize, we end this section with a "kindergarten" view of the preceding equilibrium expressions (4.32) and (4.33). Refer again to Fig. 5.7. The element of fluid is subjected to pressure forces on its sides ($p_\perp$-related) and a "buoyancy force" ($p_\parallel$-related, which can be interpreted as the differential of parallel pressure forces on the tops of the flux element), and to two internal magnetostrictive forces: the mirror force (which can be interpreted as being responsible for the "slippery soap" effect of a narrowing magnetic flux tube squeezing the incoming particles and bouncing them back in their parallel motion) and the inertial centrifugal force (on the guiding center particles in a bent flux tube while they travel up and down in their bounce motion). If this force system is in hydrostatic equilibrium, there is no macroscopic bulk acceleration in any direction (perpendicular or parallel to $\boldsymbol{B}$). As a result, the particle ensemble is stationary; a locally time-independent guiding center bulk flow $\boldsymbol{V}_{g\perp}$ is allowed, but only in such a way that no total current $\boldsymbol{J}_\perp$ occurs. No net field-aligned bulk flow (or current) is allowed.

This entire discussion involved the guiding center fluid model—the kinetic model does not care about what the individual particles do elsewhere (like whether they are executing a systematic cyclotron gyration and come back to the same volume element repeatedly to be counted each time as a contribution to a current, or whether they fly away and are replaced by other incoming particles); what counts in the kinetic fluid model is what happens *locally* to each particle at any given point in space and instant of time. In this more general kinetic formalism one loses track of the integral, macrophysical picture and related intuitive understanding. Remarkably, however, as we shall see in the next section the equations discussed above are valid also in the kinetic particle model, with the velocities and pressures defined in the list on page 124. This is a relief because, as hinted before, the kinetic fluid model is the only recourse available for the quantitative study of regions in which the adiabatic conditions break down for the particles in question, such as in the vicinity of neutral sheets and lines, boundaries and shocks.

5.4 From the Guiding Center Fluid to a Quasi-neutral Center-of-Mass Fluid

It is high time to turn to *quasi-neutral mixtures* of positive and negative plasma particles. Most of the examples to be considered will be, for simplicity, singly-charged positive ions and negative electrons. We also must turn our attention to the fact that, as hinted at the beginning of this chapter, in a real plasma we cannot neglect the contribution of the plasma currents to the field: there is a circular cause-effect relationship: particle dynamics ⇒ currents ⇒ magnetic field ⇒ particle dynamics—in other words, we must activate the link between Maxwell's equations and fluid dynamics, in which the magnetic field still plays the grand role of a common framework holding the different components and regions of a plasma together. It is important to point out, as we shall see in a later section, that neither fields nor particles come first (a "chicken-and-egg" situation)—except when one or the other has separate and dominating *externally controlled* sources or sinks (e.g., the internal geomagnetic field; solar wind particle injections, atmospheric losses). And since thus far we were dealing with collisionless ensembles in which the only interaction between particles is mediated by the macroscopic electromagnetic field, at one point we must get real and turn inter-particle collisions on.

First of all, we start with Eq. (5.18) for one species and note that it really can also be derived *directly* from Vlasov's equation (5.1) for collisionless ensembles: just multiply all terms of this equation tensorially by $m\boldsymbol{v}$ and integrate over velocity space! This means that it is valid for a kinetic fluid, too, provided one accepts the fact that, as mentioned on page 135, for the bulk accelerations $d\boldsymbol{V}_g/dt = d\boldsymbol{V}/dt$ despite both velocities being different. For that reason, we shall drop the subindex "g" from the velocity vector $\boldsymbol{V}$.[8] This means that (5.18) has more general validity than the perpendicular and parallel guiding center fluid equations from which we extracted it.

It is our task now to merge two ensembles with mutually opposite charges, each one representing a class of particles under one common electromagnetic field. In this way we obtain yet another fluid which provides a quantitative macroscopic description of the overall system, and from which one can extract some useful information about the behavior of each one of the merged ensembles. To develop this "grand" new fluid model, we shall use $+$ and $-$ as subindices characterizing each species. With this notation, we rewrite (5.18) in the forms

$$n_+ m_+ \frac{d\boldsymbol{V}_+}{dt} = n_+ q_+ \boldsymbol{E} - \nabla\mathbb{P}_+ + \boldsymbol{J}_+ \times \boldsymbol{B} \tag{5.21}$$

$$n_- m_- \frac{d\boldsymbol{V}_-}{dt} = n_- q_- \boldsymbol{E} - \nabla\mathbb{P}_- + \boldsymbol{J}_- \times \boldsymbol{B} \tag{5.22}$$

To these we must add a continuity equation for each species (we are assuming that there are no sources or sinks of particles in our collisionless mixed ensembles):

[8] A question still subsists: How can two different solutions, either $\boldsymbol{V}_g$ or $\boldsymbol{V}$, be obtained for the two different fluid models from one and the same equation? The answer is that $\boldsymbol{V}_g$ or $\boldsymbol{V}$ sit inside $\boldsymbol{J}$, which in the case of the *magnetized* guiding center fluid model also contains $\boldsymbol{\nabla} \times \boldsymbol{M}$, with $\boldsymbol{M}$ in turn being a function of $\boldsymbol{B}$ and $p_\perp$.

$$\frac{\partial n_\pm}{\partial t} + \nabla \cdot (n_\pm \boldsymbol{V}_\pm) = 0 \tag{5.23}$$

What specific properties must we expect from our new single-fluid mixed-species model? First of all, its mass, charge and current densities should be the sum of the individual densities:

$$\rho_m = n_+ m_+ + n_- m_- \tag{5.24}$$

$$\rho_q = n_+ q_+ - n_- |q_-| \tag{5.25}$$

$$\boldsymbol{J} = \boldsymbol{J}_+ + \boldsymbol{J}_- = n_+ q_+ \boldsymbol{V}_+ - n_- |q_-| \boldsymbol{V}_- \tag{5.26}$$

Second, the bulk velocity of the fluid $\boldsymbol{V}$ should be such that the momentum density $\boldsymbol{G} = \rho_m \boldsymbol{V}$ is equal to the sum of the momentum densities of each component:

$$\boldsymbol{G} = n_+ m_+ \boldsymbol{V}_+ + n_- m_- \boldsymbol{V}_- \tag{5.27}$$

For that purpose we now introduce the *center of mass velocity* of the two fluids:

$$\boldsymbol{V} = \frac{n_+ m_+ \boldsymbol{V}_+ + n_- m_- \boldsymbol{V}_-}{n_+ m_+ + n_- m_-} \tag{5.28}$$

We can now officially introduce the *center of mass fluid* as one with mass density ρ_m (5.24), charge density ρ_q (5.25), current density $\boldsymbol{J}$ (5.26) and momentum density $\boldsymbol{G} = \rho_m \boldsymbol{V}$. A continuity equation can be derived from (5.23),

$$\frac{\partial \rho_m}{\partial t} + \nabla \cdot (\rho_m \boldsymbol{V}) = 0 \tag{5.29}$$

with a charge continuity equation

$$\frac{\partial \rho_q}{\partial t} + \nabla \cdot (\rho_q \boldsymbol{V}) = 0 \tag{5.30}$$

We have sketched the situation of a center of mass fluid in Fig. 5.8:Consider an element of volume $\delta \boldsymbol{r}^3$ at time t with two classes of particles of opposite charge, ∘ and •.[9] Each class came, in principle, from a different volume element at time $t - \delta t$, and each will end up in a different parcel at time $t + \delta t$. The centers of mass of the parcel pairs at these different times are shown (please note that in reality these parcels are only infinitesimal time intervals and distances apart!). With the center of mass fluid model we have replaced two distinct, intercrossing ∘ and • fluids with

[9]Of course, we can show only a subgroup of particles of each class in the central element of volume at time t; it may be crossed by many other particles coming from other pairs of pre-t parcels.

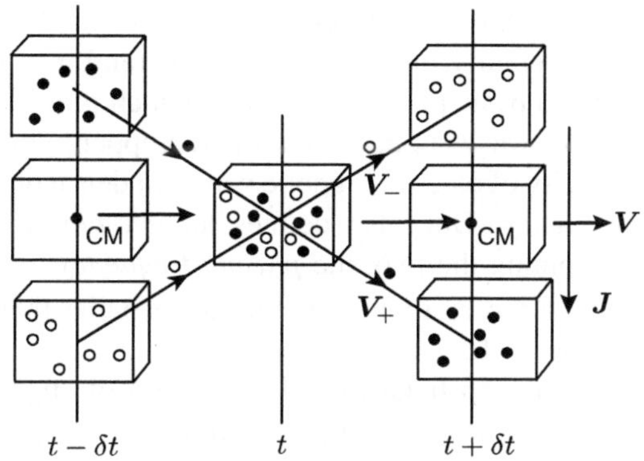

Fig. 5.8 Sketch of two guiding center fluids in their partial motions and that of the virtual center of mass fluid. The corresponding electric current density vector is shown. In an ion-electron fluid, the center of mass motion would be nearly identical with the bulk motion of the ions

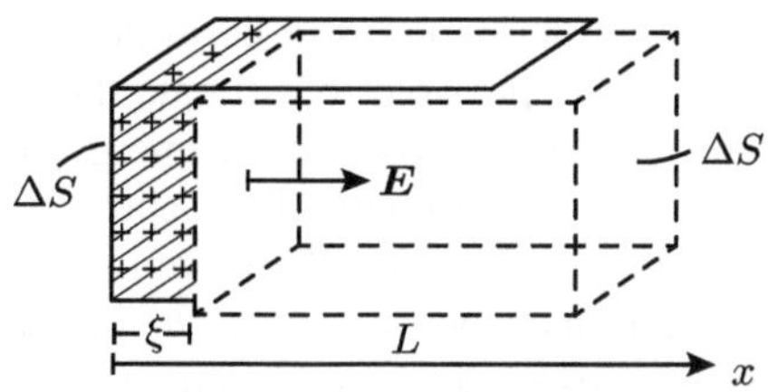

Fig. 5.9 Mutual displacement of positive and negative particle ensembles; generation of a local electrostatic field

another virtual fluid whose parcels follow the center of mass line (horizontal in the figure), whose (virtual) density is the sum of the individual fluid densities, and which sustains a current density given by the relative convection of opposite charges (vertical in the figure).

Figure 5.8 shows two ensembles of oppositely charged particles running through each other—What prevents them from separating, what keeps them together? Evidently, it must be the electric field that would build up rapidly between oppositely charged "clouds" of particles if a local charge density fluctuation occurs in an electrically neutral distribution. As a matter of fact, just a tiny collective charge separation in a limited volume would build up a space charge and generate a large local electric field (thanks to the large value of the constant ϵ_0, Appendix A.1) acting against any further separation. Refer to Fig. 5.9 showing a portion of the electron population displaced by an amount ξ to the right ($\xi \ll L$), and assume for a moment that all particles are "frozen" into their instantaneous position. A positive electrostatic space charge will appear in the thin rectangular element of volume $\Delta S\,\xi$ ($\xi^2 \ll \Delta S$).

The electric field on the right of this thin, flat element would be approximately uniform and of value $E = (ne\xi)/\epsilon_0$, directed as shown (e: absolute value of the elementary charge; $ne\xi$: charge per unit surface S of the thin element). This field will exert a total force $F = qE = -(neL\Delta S)(ne\xi)/\epsilon_0$ on the electron cloud of total mass $M = nm_e L\Delta S$. Newton's equation $F = M\ddot{\xi}$ turns out to be that of an harmonic oscillator $\ddot{\xi} + {\omega_e}^2\,\xi = 0$, where

$$ {\omega_e}^2 = \frac{ne^2}{\epsilon_0 m_e} \tag{5.31} $$

Despite the very artificial nature of the model used in the derivation, this frequency of coherently oscillating electrons (and also of the oscillating local electric field $\boldsymbol{E}$) is a basic plasma parameter which plays a fundamental role in certain types of plasma waves (e.g., Langmuir and upper hybrid waves) and affects the propagation velocity of electromagnetic waves in dense plasmas. It is called the *electron plasma frequency*. One also defines an *ion plasma frequency* ${\omega_i}^2 = ne^2/\epsilon_0 m_i$, and, combining the two, the *plasma frequency* $\omega_p^2 = \omega_e^2 + \omega_i^2$. However, since ${\omega_i}^2$ is at least three orders of magnitude smaller than ${\omega_e}^2$, one also uses the term plasma frequency for ω_e.

In our artificial model, the cloud of slightly displaced electrons will oscillate about the neutral charge position (where $\rho_e = \rho_i$) with a frequency ω_e that only depends on their number density (5.31). By loosening the initial restrictions and allowing the particles to be in thermal motion (no collisions and still no magnetic field), it is possible to estimate the upper limit of the amplitude A_{max} noting that the electrostatic oscillation energy must compete with the kinetic energy of the particles. A reasonable limit should therefore be $A_{max} \cong \langle|v|\rangle/\omega_e$, where $\langle|v|\rangle$ is the average thermal velocity of the electrons. Taking into account (5.31) and (4.22), we introduce the *Debye length* as another important plasma parameter, which in our specific example would be equal to A_{max}:

$$ {\lambda_D}^2 = \frac{\epsilon_0 k T_e}{ne^2} \tag{5.32} $$

Removing the last artificial restriction and allowing the presence of a magnetic field $\boldsymbol{B}$ will complicate the model of Fig. 5.9 considerably because of the additional action of the Lorentz force[10]; one should anticipate that the behavior will be different for a magnetic field parallel to the x-axis than for a perpendicular field. If we call ρ_{Ce} the average cyclotron radius of the electrons, we can write

$$ \frac{\omega_c}{\omega_e} = \frac{\lambda_D}{\rho_{Ce}} = \frac{1}{c}\frac{B}{\sqrt{(\mu_0 n m_e)}} \tag{5.33} $$

The lower limit of this ratio in the magnetosphere is about 0.001; only in some very limited regions it may exceed 1. This is an indication that the electrostatic effect in our simplified model sketched in Fig. 5.9 will in general affect only a small portion of a typical Larmor orbit in the magnetosphere; the "thermal" motion is then given by the random cyclotron phases (even for mono-energetic particles)—neglecting the magnetic field in the derivation of (5.32) was not such an unrealistic choice after all!

[10]In most plasma physics books, the Debye length and the plasma frequency are introduced at the very beginning, without any mention of the magnetic field (and often assuming a Maxwellian distribution). This sometimes confuses the student, especially if the book mainly deals with *magnetized* plasmas.

The requirement that $\lambda_D \ll$ scale of the system is usually taken as the very *definition of a plasma*. For what follows, we need not concern ourselves with the detailed mechanism by means of which quasi-neutrality $\rho^+ \simeq \rho^-$ is maintained; we simply make the assumption that this condition is upheld at all times through some local mechanism. This is quite similar to what one does in analytical mechanics: setting predetermined constraints without in any way specifying how such limiting conditions are physically maintained!

Returning to our center of mass fluid, it is important to be aware of the relationships between the species' bulk velocities and the total current:

$$V_\pm = \boldsymbol{V} + \frac{m_\mp}{q_\pm} \boldsymbol{J}/\rho_m \tag{5.34}$$

Carefully note the $+$ and $-$ correspondences. If the negative particles are electrons, and the ensemble is quasi-neutral, we can write:

$$\begin{aligned}
n &\simeq n^+ \simeq n^- \\
\rho_m &\simeq n_+ m_+ \\
\rho_q &\simeq 0 \\
\boldsymbol{J} &= \boldsymbol{J}_+ + \boldsymbol{J}_- \simeq ne(\boldsymbol{V}_+ - \boldsymbol{V}_-) \simeq -ne\boldsymbol{V}_- \\
\boldsymbol{V}_+ &\simeq \boldsymbol{V} \;\text{ and }\; \boldsymbol{V}_- \simeq \boldsymbol{V} - \boldsymbol{J}/ne
\end{aligned} \tag{5.35}$$

We now have to come up with a single dynamic equation for the center of mass fluid. Unfortunately, we cannot simply add algebraically the species-specific equations (5.21) and (5.22). There are several reasons. First, they contain total time derivatives, which follow each species of particles traveling with different bulk speeds in different directions. So we need to break them up into local and convective operators: $d/dt = \partial/\partial t + \nabla \cdot \boldsymbol{V}_\pm$. Second, the pressure tensor components are not additive either: according to the definition (4.16) and the discussion in Sect. 4.3, the pressure tensor is controlled by the velocity dispersion of a species of particles in a frame of reference moving at each point with the bulk velocity $\boldsymbol{V}_\pm$ of *that* particular species. It is thus necessary to introduce another entity, a pressure tensor which for each species links the particles' velocity distribution with the common center of mass frame of reference. For that purpose the *partial pressure* tensor $\mathbb{P}^*_\pm$ is introduced, defined as

$$\mathbb{P}^*_\pm = m_\pm \int f_\pm (\boldsymbol{v} - \boldsymbol{V}) \otimes (\boldsymbol{v} - \boldsymbol{V}) d\boldsymbol{v}^3 \tag{5.36}$$

in which $\boldsymbol{V}$ now is the center of mass velocity (5.28). Note the algebraic relationship with the respective species-specific kinetic tensors (4.15) which indeed *are* additive:

$$\mathbb{P}^*_\pm = \mathbb{K}_\pm - n_\pm m_\pm \boldsymbol{V} \otimes \boldsymbol{V} \tag{5.37}$$

Therefore, the partial pressure tensors $\mathbb{P}^*_\pm$ are legitimately additive for different species, too.

To accomplish all this, let us go back to Eqs. (5.21) and (5.22) and replace the original pressure tensors $\mathbb{P}_\pm$ with their relation to their kinetic tensors and species-specific bulk velocities: $\mathbb{P}_\pm = \mathbb{K}_\pm - n_\pm m_\pm \boldsymbol{V}_\pm \otimes \boldsymbol{V}_\pm$ (4.16). We then have

$$\nabla \mathbb{P}_\pm = \nabla \mathbb{K}_\pm - \nabla (n_\pm m_\pm \boldsymbol{V}_\pm \otimes \boldsymbol{V}_\pm)$$

One can verify by components that

$$\nabla (n_\pm m_\pm \boldsymbol{V}_\pm \otimes \boldsymbol{V}_\pm) = n_\pm m_\pm (\boldsymbol{V}_\pm \cdot \nabla)\boldsymbol{V}_\pm + \boldsymbol{V}_\pm [\nabla \cdot (n_\pm m_\pm \boldsymbol{V}_\pm)]$$

Using the continuity equation (5.23) for the last term, we obtain

$$\nabla \mathbb{P}_\pm = \nabla \mathbb{K}_\pm - n_\pm m_\pm (\boldsymbol{V}_\pm \cdot \nabla)\boldsymbol{V}_\pm + \boldsymbol{V}_\pm \frac{\partial (n_\pm m_\pm)}{\partial t}$$

Inserting in (5.21) and (5.22), rearranging terms and remembering that $d/dt = \partial/\partial t + \boldsymbol{V} \cdot \nabla$,

$$\begin{aligned} \frac{\partial}{\partial t}(n_\pm m_\pm \boldsymbol{V}_\pm) &= n_\pm m_\pm \frac{d\boldsymbol{V}_\pm}{dt} - n_\pm m_\pm (\boldsymbol{V}_\pm \cdot \nabla)\boldsymbol{V}_\pm + \boldsymbol{V}_\pm \frac{\partial (n_\pm m_\pm)}{\partial t} \\ &= n_\pm q_\pm \boldsymbol{E} - \nabla \mathbb{K}_\pm + \boldsymbol{J}_\pm \times \boldsymbol{B} \end{aligned}$$

With $\boldsymbol{G}_\pm = n_\pm m_\pm \boldsymbol{V}_\pm$ as the momentum density of each partial fluid, we finally have a pair of momentum equations for the two fluids which are indeed summable:

$$\begin{aligned} \frac{\partial \boldsymbol{G}_+}{\partial t} &= n_+ q_+ \boldsymbol{E} - \nabla \mathbb{K}_+ + \boldsymbol{J}_+ \times \boldsymbol{B} \\ \frac{\partial \boldsymbol{G}_-}{\partial t} &= n_- q_- \boldsymbol{E} - \nabla \mathbb{K}_- + \boldsymbol{J}_- \times \boldsymbol{B} \end{aligned} \tag{5.38}$$

Adding the two equations, we obtain

$$\frac{\partial \boldsymbol{G}}{\partial t} = \rho_q \boldsymbol{E} - \nabla \mathbb{K} + \boldsymbol{J} \times \boldsymbol{B} \tag{5.39}$$

with $\mathbb{K} = \mathbb{K}_+ + \mathbb{K}_-$ the total kinetic tensor of the center of mass fluid. This is the *momentum* magnetohydrodynamic equation.

Now we can revert to a true dynamic equation, with a total time derivative that represents the acceleration of a fluid element in the new model as it flows. This can be done by starting with (5.39) and "undoing" some of the previous steps, to obtain the familiar *magnetohydrodynamic equation*

$$\rho_m \frac{d\boldsymbol{V}}{dt} = \rho_q \boldsymbol{E} - \nabla\mathbb{P} + \boldsymbol{J} \times \boldsymbol{B} \tag{5.40}$$

Note that it looks just as the Eq. (5.18) for one species, but here $\boldsymbol{V}$ is the *center of mass velocity* and $\mathbb{P} = \mathbb{P}_+^* + \mathbb{P}_-^*$ is the *total pressure tensor* (5.36), sum of partial pressure tensors (5.37).

At once, with this equation we can retrieve several earlier relations that we have deduced for conditions of stationary equilibrium. In particular, for a given static magnetic field and $\boldsymbol{V}$ = const. there are strong restrictions on the admissible particle distributions of a quasi-neutral ensemble. Remembering relations (A.37) and (A.38) of Appendix A.1, we can write (5.40) in the form $\nabla(\mathbb{P} - \mathbb{S}) = 0$ or, in general, the equilibrium between the plasma pressure tensor and Maxwell's magnetic stress tensor $\mathbb{P} = \mathbb{S}$.

It is easy to extend the center of mass fluid equations to a mixture of particles with more components than two: if we replace the + and − subindices in the preceding derivations with the subindex s, we just have to sum everything over s. We end up with the following list of macroscopic variables for a center of mass fluid of any number of constituents:

Total mass density: $\rho_m = \sum_s n_s m_s$
Total charge density: $\rho_q = \sum_s n_s q_s$ ($\simeq 0$ in quasi-neutrality)
Total current density: $\sum_s n_s q_s \boldsymbol{V}_s$
Bulk or center of mass velocity: $\boldsymbol{V} = (\sum_s n_s m_s \boldsymbol{V}_s)/(\sum_s n_s m_s)$
Total momentum density: $\boldsymbol{G} = \rho_m \boldsymbol{V}$
Total pressure tensor: $\mathbb{P} = \int \sum_s m_s f_s(\boldsymbol{v} - \boldsymbol{V}) \otimes (\boldsymbol{v} - \boldsymbol{V}) d\boldsymbol{v}^3$ (sum of partial pressure tensors).

With these macroscopic variables, the continuity equation (5.29) remains unchanged, and so does the magnetohydrodynamic equation (5.40).

5.5 Collisions and the Generalized Ohm Equation

It is prudent to take stock of what we have accomplished so far in the development of quantitative relationships between macroscopic variables for quasi-neutral ensembles of electrically charged particles, and the dynamic equations governing their time changes. Electrostatic forces which may appear on a mesoscopic scale (the Debye length (5.32), large compared to inter-particle distances but small with respect to the overall scale of the system) overwhelm the local magnetic field forces that normally dominate the behavior of a collisionless particle ensemble and prevent any local charge density fluctuations from growing to a macroscopic scale. This omnipresent mechanism justifies adopting quasi-neutrality as one of the defining properties of a plasma.

To arrive at the momentum and magnetohydrodynamic equations we have followed two possible routes by introducing two models. (1) The model of a guiding center fluid ruled by adiabatic theory, restricted to situations in which the guiding center approximation, gyrotropicity and trapping conditions are valid at all points of the fluid. (2) A particle or kinetic fluid model that follows directly from the Vlasov equation (5.1), with no restrictions such as adiabatic conditions. The latter, however, does not offer the intuitive visualization of "what particles are really doing at the microscopic level, and why". Both approaches lead to identical results wherever the adiabatic conditions are satisfied, and both involve a distribution function (of six and seven variables, respectively) as the fundamental physical and measurable quantity at the mesoscopic level. Initially formulated for just one species of particles, we combined two oppositely charged species into a quasi-neutral mixture by introducing yet another model, the center of mass fluid. As a fundamental result we obtained a single center of mass fluid momentum equation (5.39) and the magnetohydrodynamic (MHD) equation (5.40).

The principal aim of this formalism is to be able to predict or retrodict the behavior of a given plasma, eventually subjected to some externally controlled electromagnetic field and particle sources and sinks, which at a given initial time is found in a given macroscopic state. In more practical terms for magnetospheric physics, the aim is to develop a mathematical framework that, given some observed large-scale phenomena such as the trigger and development of a magnetospheric substorm, an auroral breakup, a sudden energetic trapped particle injection, etc., would allow us to pinpoint the ultimate external cause, understand the quantitative evolution, and formulate associated prediction algorithms. Taken in isolation, the MHD equation would be useful *only* to address some oversimplified situations, where there is an a priori set symmetries and isotropies, absence of collisions, and a priori imposed equilibrium conditions. In fact, to use it at this stage, the electromagnetic field vectors must be *pre-specified*, and all retro-effects on the plasma on the field must be ignored. The real problem is that we still have too many unknowns but not enough equations: we have not yet properly *linked* the MHD equation with the overall electromagnetic field!

The MHD equation was derived by manipulating the two momentum equations (5.38) for oppositely charges species. From the mathematical point of view, we are still allowed to extract one more independent equation from those two, which explicitly reflects the local interaction between plasma and field. But before we do so, we shall introduce elastic collision processes (Coulomb scattering) between the particles of the ensemble, thus dropping yet another of the restrictions imposed at the beginning of this chapter. Let us call $\boldsymbol{k}_{rs}$ the average *momentum transfer density* per unit time from the fluid of s particles to the r-particle species. Obviously, $\boldsymbol{k}_{rs} = -\boldsymbol{k}_{sr}$ for elastic collisions. These quantities would then have to be added, respectively, to each momentum equation (we assume that although particles may collide with their own kind, there should be no net average momentum transfer between them). It should be clear, then, that this addition would not affect at all the procedure followed on page 146 (the extra collision terms would cancel each other), which means that the MHD equation (5.40) *is valid even in presence of elastic collisions.* For our next purpose, however, we have to come up with a quantitative expression for $\boldsymbol{k}$; to simplify the argument, we shall do it for a quasi-neutral mixture of singly-charged positive ions and electrons. If $\boldsymbol{V}_i$ and $\boldsymbol{V}_e$ are the average bulk velocities of ions and electrons, it is reasonable to assume that $\boldsymbol{k}_{ie} = C(\boldsymbol{V}_e - \boldsymbol{V}_i)$, where C should be proportional, again on the average, to the mass density of electrons times their collision frequency ν_{coll} with ions: $n_e m_e \nu_{coll}$.[11] And, for Coulomb interactions, it *also* should be proportional on the average to the absolute value of their charge densities $e^2 n_i n_e$, which for charge neutrality amounts to $e^2 n^2$. In summary, we can set

$$\boldsymbol{k}_{ie} = \eta\, e^2 n^2 (\boldsymbol{V}_e - \boldsymbol{V}_i) \text{ with } \eta = \frac{1}{\sigma} = \nu_{coll}\frac{m_e}{e^2 n} = \frac{\nu_{coll}}{\epsilon_0 \omega_e^2} \tag{5.41}$$

ω_e is the electron plasma frequency (5.31), η is called the plasma *resistivity* and σ its conductivity.

We return to the momentum equation as it appears in (5.38) and write it for a generic species s, adding the collision momentum exchange term $\boldsymbol{k}_s$, which now represents the total momentum transfer density to fluid s from collisions with *all other* constituents. Multiplying all terms by q_s/m_s, adding over all s and rearranging terms, we obtain, remembering the expression (4.15) for the kinetic tensor $\mathbb{K}$ and the multispecies expression for the current density on page 147:

$$\frac{\partial \boldsymbol{J}}{\partial t} + \nabla \int (\sum_s q_s f_s) \boldsymbol{v} \otimes \boldsymbol{v}\, d\boldsymbol{v}^3$$
$$= \sum q_s{}^2 n_s/m_s\, \boldsymbol{E} + (\sum q_s{}^2 n_s/m_s \boldsymbol{V}_s) \times \boldsymbol{B} + \sum (q_s/m_s)\boldsymbol{k}_s \tag{5.42}$$

The integral can be re-written by considering the following relation involving the partial pressures $\mathbb{P}_s^*$ (5.36):

$$\sum (q_s/m_s)\mathbb{P}_s^* = \int \sum_s q_s f_s (\boldsymbol{v} - \boldsymbol{V}) \otimes (\boldsymbol{v} - \boldsymbol{V})\, d\boldsymbol{v}^3$$
$$= \int \sum_s q_s f_s \boldsymbol{v} \otimes \boldsymbol{v}\, d\boldsymbol{v}^3 + \rho_q \boldsymbol{V} \otimes \boldsymbol{V} - \boldsymbol{V} \otimes \boldsymbol{J} - \boldsymbol{J} \otimes \boldsymbol{V}$$

This leads us to the weird-looking *generalized Ohm equation*:

$$\frac{\partial \boldsymbol{J}}{\partial t} + \nabla\Big(\boldsymbol{V} \otimes \boldsymbol{J} + \boldsymbol{J} \otimes \boldsymbol{V} - \rho_q \boldsymbol{V} \otimes \boldsymbol{V}\Big) \tag{5.43}$$
$$= \Big(\sum q_s{}^2 n_s/m_s\Big) \boldsymbol{E} + \Big(\sum q_s{}^2 n_s/m_s \boldsymbol{V}_s\Big) \times \boldsymbol{B} - \nabla \sum q_s/m_s \mathbb{P}_s^* + \sum (q_s/m_s)\boldsymbol{k}_s$$

---

[11]It is assumed that the actual momentum transfer can vary with equal probability distribution between 0 and a maximum of $2 n_e m_e \nu_{coll}$.

What we have here is a companion equation to the dynamic fluid equation (5.40), connecting field sources $\boldsymbol{J}$ and ρ_q *internal* to the center of mass fluid to the macroscopic particle ensemble variables $\boldsymbol{V}, n_s, \mathbb{P}_s^*$ and $\boldsymbol{k}_s$. This connection is *local*, but the overall connecting agent is the electromagnetic $\boldsymbol{E}$ and $\boldsymbol{B}$ field, which, as discussed in Appendix A.1, depends on *all* charges and currents, including those externally controlled which have nothing to do with the plasma under consideration. This means that to Eqs. (5.40) and (5.43) we must add Maxwell's equations (Appendix A.1, (A.48)–(A.51)) in which the current density $\boldsymbol{J}$ (and the charge density ρ_q) must also include *all external sources*, and complete the set with the conservation equations (5.29) and (5.30).[12] Note the distinct character of each one: (1) The MHD equation controls the dynamics of the particle ensemble—it must be integrated like any dynamics equation to provide information on temporal behavior. (2) The generalized Ohm equation binds together local properties of plasma and field—there is nothing there to integrate, but it leads to the electric field which then appears in the Eq. (A.53) defining $\partial \boldsymbol{B}/\partial t$. (3) Maxwell's equations tie together concurrent behavior at distant points—concurrent in the relativistic sense (however, retardation ((A.41) and (A.42)) usually plays no role in plasmas of planetary system dimension). (4) Conservation equations are the "balance sheets" for the movements of mass and electric charge.

The resulting equation framework is, unfortunately, unmanageable, and we must first trim some fat from Ohm's general equation before we can turn to some simple examples. Instead of first doing a rigorous comparative analysis of the order of magnitude of different terms under different conditions, our first step will be to again limit ourselves to electrons and singly charged positive ions, under guaranteed quasi-neutrality $n_i = n_e = n$, $\rho_q = n(q_i + q_e) = 0$ (the term "quasi" meaning eventual allowance for little departures from charge neutrality within a Debye domain). Under these conditions, we have the following relations for some of the coefficients in Eq. (5.43):

$$\sum q_s{}^2 n_s/m_s = e^2 n \frac{m_i + m_e}{m_i m_e} = \epsilon_0(\omega_e^2 + \omega_i^2)$$

$$\sum (q_s/m_s)\boldsymbol{k}_s = \eta e^3 n^2 \frac{m_i + m_e}{m_i m_e}(\boldsymbol{V}_i - \boldsymbol{V}_e) = e^2 n \frac{m_i + m_e}{m_i m_e}\eta \boldsymbol{J}$$

and

$$\sum q_s{}^2 n_s/m_s \boldsymbol{V}_s = e^2 n \frac{m_e \boldsymbol{V}_i + m_i \boldsymbol{V}_e}{m_i m_e}$$

We took into account relations (5.31) and (5.41). Multiplying the generalized Ohm equation for a two-component plasma by the first factor above, we obtain:

[12] Equation (5.43) only includes plasma-driven currents (5.9) and (5.15)—herein lies the crux of understanding correctly the "chicken-and-egg" question of what comes first, $\boldsymbol{B}$ or $\boldsymbol{J}$?! See also [1]. The set of equations (5.40), (5.43), (5.29) and (5.30) is usually called *the MHD equations* (plural!).

$$\frac{m_i m_e}{m_i + m_e} \frac{1}{n\, e^2} \left[\frac{\partial \boldsymbol{J}}{\partial t} + \nabla(\boldsymbol{V} \otimes \boldsymbol{J} + \boldsymbol{J} \otimes \boldsymbol{V}) \right]$$
$$= \boldsymbol{E} + \boldsymbol{V} \times \boldsymbol{B} - \eta \boldsymbol{J} - \frac{m_i - m_e}{m_i + m_e} \frac{\boldsymbol{J} \times \boldsymbol{B}}{n\, e}$$
$$- \frac{1}{(m_i + m_e)} \frac{1}{ne} \nabla\left(m_e \mathbb{P}_i^* - m_i \mathbb{P}_e^*\right) \tag{5.44}$$

For isotropic pressures p_i and p_e, the divergence vectors of the partial pressure tensors become gradient vectors of the respective scalar pressures. As an aside, note that if instead of ions and electrons we had a positron-electron or an antiproton-proton plasma ($m_+ = m_-$), the last two terms in the right side would drop out, and we would be left with a very simple generalized Ohm equation. Unfortunately, it is too dangerous to play with such plasmas, especially if $\eta \neq 0$, so our next great simplification will rather be to stick to an ion-electron plasma and take into account that $m_e \ll m_i$. Hence, $\boldsymbol{V} \simeq \boldsymbol{V}_i$, which leads to the following equation (using (5.31)):

$$\frac{1}{\epsilon_0 \omega_p^2} \left[\frac{\partial \boldsymbol{J}}{\partial t} + \nabla(\boldsymbol{V} \otimes \boldsymbol{J} + \boldsymbol{J} \otimes \boldsymbol{V}) \right]$$
$$= \boldsymbol{E} + \boldsymbol{V} \times \boldsymbol{B} - \eta \boldsymbol{J} - \frac{\boldsymbol{J} \times \boldsymbol{B}}{n\, e} + \frac{1}{n\, e} \nabla\left(\mathbb{P}_e^*\right) \tag{5.45}$$

For an ion-electron plasma, this reduced Ohm equation has a basic physical interpretation, namely, that it is equivalent to a dynamic equation for the electron fluid (with a collision term) *as seen from a reference system fixed to, and traveling with the ion fluid.* In other words, we can imagine the ion-electron plasma as a mass fluid (the ions) and, embedded in it, a massless negative charge fluid (the electrons) guided by its own dynamic equation, and whose flow confers the main electromagnetic properties to the coupled system. The transformation of Eq. (5.45) from the original frame of reference to the ion fluid frame is simple but lengthy [2]; here we will just ask ourselves how such an equation would look. For that purpose, we start with an equation of the type (5.18) (plus the collision term). Calling $\boldsymbol{V}^* = \boldsymbol{V}_e - \boldsymbol{V}_i$ the velocity of the electron fluid, we have $\boldsymbol{J} = ne\boldsymbol{V}^*$ and $\boldsymbol{E}^* = \boldsymbol{E} + \boldsymbol{V}_i \times \boldsymbol{B}$ the electric field seen in the frame moving with the ion fluid, the equation in the moving ion frame should obviously be

$$n_e\, m_e \frac{d\boldsymbol{V}^*}{dt} = -\, n_e e(\boldsymbol{E}^* + \boldsymbol{V}^* \times B) - \nabla \mathbb{P}_e^* - \eta n_e^2 e^2 \boldsymbol{V}^* - n_e\, m_e (\boldsymbol{V}^* \cdot \nabla) \boldsymbol{V}_i$$

The additional last term is the *inertial force* density acting on the electron fluid due to the acceleration of the frame of reference used (motional change in velocity $\boldsymbol{V}_i$). Transforming the relevant quantities back to the original frame of reference, leads indeed to Eq. (5.45)! One might argue whether this invalidates the earlier assertion that the generalized Ohm equation is not a dynamic equation, but one which brings out local relations in the center of mass fluid. It does not, because although it is a dynamic equation in the ion frame, it describes only a *part* of the whole system.

We may now venture to discuss some simple examples. First we shall examine Eq. (5.45), neglecting the left side on the grounds that it is a quantity divided by the square of the electron plasma frequency. We then can write for the natural system components of the electric field *in the center of mass fluid*:

$$\begin{aligned} \boldsymbol{E}_{\perp}^{cm} &= \boldsymbol{E}_{\perp} + \boldsymbol{V} \times \boldsymbol{B} = \eta \boldsymbol{J}_{\perp} + \frac{\boldsymbol{J} \times \boldsymbol{B}}{ne} - \frac{\nabla_{\perp}\mathbb{P}_e^*}{ne} \\ \boldsymbol{E}_{\parallel}^{cm} &= \boldsymbol{E}_{\parallel} = \eta \boldsymbol{J}_{\parallel} - \frac{\nabla_{\parallel}\mathbb{P}_e^*}{ne} \end{aligned} \tag{5.46}$$

The quantity $\boldsymbol{E}_{\perp}^{cm}$ is the electric field seen in the center of mass fluid; the first term on the right side represents the *ohmic resistance field*; the second term is the *Hall field* (which exists in *any* current-carrying conductor placed in a magnetic field); and the third term is called the *ambipolar electric field* (similar to the field responsible for the e.m.f. in a battery). We now turn to the MHD equation, which in stationary state (and charge neutrality) leads to $\boldsymbol{J} \times \boldsymbol{B} = \nabla_{\perp}\mathbb{P}$ ($\mathbb{P} = \mathbb{P}_i^* + \mathbb{P}_e^*$). Taking this into account, the transverse component in (5.46) becomes

$$\boldsymbol{E}_{\perp} + \boldsymbol{V} \times \boldsymbol{B} = \eta \boldsymbol{J}_{\perp} - \frac{\nabla_{\perp}\mathbb{P}_i^*}{ne} \tag{5.47}$$

Multiplying vectorially by $\boldsymbol{B}/B^2$, we obtain an expression for the perpendicular component of the bulk velocity of the center of mass fluid (nearly equal to that of the ion fluid):

$$\boldsymbol{V}_{\perp} = \frac{\boldsymbol{E} \times \boldsymbol{B}}{B^2} - \frac{\eta}{B^2}\nabla_{\perp}\mathbb{P} - \frac{\nabla\mathbb{P}_i^* \times \boldsymbol{B}}{neB^2} \tag{5.48}$$

The first term is the pure electric drift velocity—the velocity a near-zero energy probe particle would have, and therefore also the velocity of a magnetic field line (1.38), provided no potential electric fields are present. The second term (with its sign) is called *diffusion velocity* and the third term (also with its sign) is the *diamagnetic ion drift velocity* (think of the surface currents on the cylinders in Fig. 5.2!).

Next, we shall neglect the term containing the vector divergence of the total pressure tensor divided by the number density n in (5.46). This leaves us with the following pair for the electric field natural components in the center of mass fluid:

$$\begin{aligned} \boldsymbol{E}_{\perp}^{cm} &= \boldsymbol{E}_{\perp} + \boldsymbol{V} \times \boldsymbol{B} = \eta \boldsymbol{J}_{\perp} + \frac{\boldsymbol{J} \times \boldsymbol{B}}{ne} \\ \boldsymbol{E}_{\perp}^{cm} &= \eta \boldsymbol{J}_{\parallel} \end{aligned} \tag{5.49}$$

With a little vector algebra we arrive at the following:

$$\boldsymbol{E}^{cm} \times \boldsymbol{B} = \eta ne \boldsymbol{E}_{\perp}^{cm} - \eta^2 ne \boldsymbol{J}_{\perp} - \boldsymbol{J}_{\perp} \frac{B^2}{ne}$$

or

$$\boldsymbol{J}_{\perp} = \frac{\eta ne \boldsymbol{E}_{\perp}^{cm} - \boldsymbol{E}^{cm} \times \boldsymbol{B}}{B^2/ne + \eta^2 ne}$$

We now introduce a series of plasma parameters, particularly important in ionospheric physics:

Field-aligned conductivity: $\sigma_{\parallel} = \sigma = 1/\eta$
Hall coefficient (take into account (1.21) and (5.41):

$$H = \frac{\omega_C}{\nu_{ei}} = \frac{B}{\eta ne} \tag{5.50}$$

Transverse conductivity: $\sigma_T = \sigma/(1 + H^2)$
Hall conductivity: $\sigma_H = \sigma/(H + 1/H)$

With these designations, the expression for the perpendicular component of the current density becomes:

$$\boldsymbol{J}_{\perp} = \sigma_T \boldsymbol{E}_{\perp}^{cm} + \sigma_H \boldsymbol{e} \times \boldsymbol{E}^{cm} \tag{5.51}$$

All this, including the parallel equation, can be condensed into one tensor equation

$$\boldsymbol{J} = \boldsymbol{\sigma} \boldsymbol{E}^{cm} \tag{5.52}$$

where $\boldsymbol{\sigma}$ is the grand *conductivity tensor*[13]:

$$\boldsymbol{\sigma} = \begin{pmatrix} \sigma_T & -\sigma_H & 0 \\ \sigma_H & \sigma_T & 0 \\ 0 & 0 & \sigma_{\parallel} \end{pmatrix}$$

It is important to note that even in absence of collisions ($\eta = 0$), there is a relation between the current density and the electric field in the center of mass system: $\boldsymbol{J}_{\perp} = ne/B\boldsymbol{e} \times \boldsymbol{E}^{cm}$. This is why $\boldsymbol{E}^{cm}$ is also called the *Hall field*. In this case of zero resistivity, and considering that $\boldsymbol{V} \simeq \boldsymbol{V}_i$, we can conclude from the second equality in (5.49) that

$$0 = \boldsymbol{E}_{\perp} + \boldsymbol{V}_i \times \boldsymbol{B} - (\boldsymbol{V}_i - \boldsymbol{V}_e) \times \boldsymbol{B} = \boldsymbol{E}_{\perp} + \boldsymbol{V}_e \times \boldsymbol{B} \tag{5.53}$$

---

[13]Radio propagation engineers define the Hall coefficient and Hall conductivity with *opposite sign*.

But this just tells us that the electric field in the electron fluid is zero in this case! In other words, a near-zero probe particle in the OFR will drift *with* the electron fluid. Calling up the image of moving field lines, in a collisionless plasma *magnetic field lines are "frozen" into the electron fluid.*

Our next example will have a further simplification in the reduced Ohm equation (5.45): not only will we neglect the left-hand side, but by comparing the order of magnitude of the Hall term ($[JB/ne] = [H\eta J]$, see (5.50) and (5.41)) with that of the resistive term $[\eta J]$, we see that it, too, can be neglected when the Hall coefficient H is sufficiently small (resistivity sufficiently high). And if we neglect the Hall term, we can also neglect $\nabla \mathbb{P}/ne$, because if we assume a stationary state $dV/dt = 0$, we have $\boldsymbol{J} \times \boldsymbol{B} = \nabla \mathbb{P}$. In the Maxwell equations, we shall consider $\partial \boldsymbol{E}/\partial t = 0$, and that there are *no external sources* of the field. These "fat-cutting" measures leave us with the following set of equations:

$$
\begin{aligned}
\boldsymbol{E} &= \eta \boldsymbol{J} - \boldsymbol{V} \times \boldsymbol{B} \\
\nabla \times \boldsymbol{B} &= \mu_0 \boldsymbol{J} \\
\nabla \times \boldsymbol{E} &= -\frac{\partial \boldsymbol{B}}{\partial t}
\end{aligned}
\tag{5.54}
$$

Of course, we always must consider $\nabla \cdot \boldsymbol{B} = 0$, $\nabla \cdot \boldsymbol{B} = 0$ and $\nabla \cdot \boldsymbol{E} = \rho_q/\epsilon_0 \simeq 0$. Inserting $\boldsymbol{E}$ into the last equation and taking into account that $\nabla\times(\nabla\times\boldsymbol{B}) = -\nabla^2\boldsymbol{B}$, we obtain the following partial differential equation:

$$
\frac{\partial \boldsymbol{B}}{\partial t} = \frac{\eta}{\mu_0}\nabla^2 \boldsymbol{B} + \nabla \times (\boldsymbol{V} \times \boldsymbol{B}) \tag{5.55}
$$

This equation tells us that, under these simplified conditions, the magnetic field in a resistive plasma can change *locally* in time because of the local resistivity *and* because of the local hydrodynamic flow pattern. It is a relationship that, by the way, is valid for all conducting fluids!

When the plasma is at rest ($\boldsymbol{V} \simeq 0$), (5.55) becomes a regular *diffusion equation*, with solutions that have a factor $e^{-t/\tau}$, with a decay time $\tau = \mu_0 L^2/\eta$ (L: scale size of the system). In absence of any external sources, the self-generated magnetic field of a plasma will decay exponentially. Since in Appendix A.1 we have assigned primary physical "reality" to the currents that sustain a magnetic field, we should re-state this: in a plasma under these conditions, the *currents* $\nabla \times \boldsymbol{B}$ will decay exponentially! The physical reason is easy to understand: collisions destroy the adiabatic behavior of the electrons; they diffuse and "smear out" the equivalent currents. Consider Fig. 5.2, and suppose that instead of sparsely populated by cycling particles, the cylinder is filled with denser, colliding ions and electrons. The boundary equivalent currents $\nabla\times\boldsymbol{M}$ are mainly carried by electrons. Collisions with the ions will disperse them and decrease exponentially the boundary surface current system. Since there is an overall uniform external field which originally was reduced inside the cylinder (diamagnetic effect), the total field intensity inside will increase back to the external field value. In the picture of moving field lines, originally outward-displaced field lines will straighten and move back into the cylinder.

The case of zero resistivity (collisionless plasma) in (5.55) should be reconsidered from the beginning. We arrived at this equation by neglecting the Hall term when compared to $\eta \boldsymbol{J}$. When the latter is zero, we must compare the Hall term with $\boldsymbol{V} \times \boldsymbol{B}$ in the simplified Ohm's equation. The current density is $\boldsymbol{J} = -ne(\boldsymbol{V}_e - \boldsymbol{V}_i)$; if $\boldsymbol{V} \simeq \boldsymbol{V}_i \gg (\boldsymbol{V}_e - \boldsymbol{V}_i)$ we can neglect the Hall term in a collisionless plasma, and (5.55) becomes

$$\frac{\partial B}{\partial t} = \nabla \times (\boldsymbol{V} \times \boldsymbol{B}) \tag{5.56}$$

This equation tells us that magnetic field flux tubes will move with the guiding center fluid. Indeed, the time-change of the magnetic flux through a contour whose points move with the fluid will be

$$\frac{d\Phi}{dt} = \int_S \frac{\partial \boldsymbol{B}}{\partial t} \cdot dA + \oint \boldsymbol{B} \cdot (\boldsymbol{V} \times \boldsymbol{dl}) = \int_S \left[\frac{\partial \boldsymbol{B}}{\partial t} - \nabla \times (\boldsymbol{V} \times \boldsymbol{B}) \right] \cdot \boldsymbol{dA} = 0$$

It can be easily shown that *any* contour on a flux tube will conserve the enclosed flux while moving with the plasma, thus preserving the identity of the entire flux tube. This was the feature that led Alfvén to formulate the concept of "frozen-in magnetic fields".

Speaking of Alfvén, we come to the last example and with it to the end of this chapter (and the book). It describes a wave process in collisionless plasmas, alluded to earlier in Chap. 4; however, we will only mention the most basic concepts (plasma waves deserve an entire book!) Consider a collisionless plasma in equilibrium in a uniform magnetic field $\boldsymbol{B}_0$ directed along the z-axis under the same conditions as in the previous paragraph ($\boldsymbol{V} = 0; \rho_m = 0$). We introduce a small perturbation $\boldsymbol{v}$ and $\boldsymbol{b}$ perpendicular to the uniform magnetic field. The equations to be used for the perturbations will be to first order in the perturbations:

$$\begin{aligned} \rho_m \frac{\partial \boldsymbol{v}}{\partial t} &= \frac{1}{\mu_0} (\nabla \times \boldsymbol{b}) \times \boldsymbol{B}_0 \\ \frac{\partial \boldsymbol{b}}{\partial t} &= (\boldsymbol{B}_0 \cdot \nabla) \boldsymbol{v} \end{aligned} \tag{5.57}$$

If $\boldsymbol{b}$ is directed along the x-axis, $\boldsymbol{v}$ will also be directed along that axis and we can find two solutions $b = b(z)$ and $v = v(z)$ that obey

$$\rho_m \frac{\partial v}{\partial t} = \frac{B_0}{\mu_0} \frac{\partial b}{\partial z}$$

and

$$\frac{\partial b}{\partial t} = B_0 \frac{\partial v}{\partial z}$$

Taking $\partial/\partial t$ of both we end up with two wave equations for v and b:

$$\left[\frac{\partial^2}{\partial z^2} - \frac{1}{V_A^2}\frac{\partial^2}{\partial t^2}\right](v; b) = 0$$

with

$$V_A = \frac{B_0}{\sqrt{\mu_0 \rho_m}} \tag{5.58}$$

the *Alfvén velocity*, with which any perturbation propagates in a collisionless plasma in the direction of the originally unperturbed magnetic field. It can be shown geometrically that in this process the elicited displacement pattern of the field line is proportional to the pattern of field and velocity perturbation and propagates with them. Alfvén waves are commonly interpreted as transverse oscillations of magnetic field lines. A bit more precisely, Alfvén waves are transverse fluid oscillations which, as they propagate in the direction of the main field, distort the field lines in their oscillatory motion. The oscillatory electric field is perpendicular to the magnetic field variations, and both are mutually out of phase by $\pi/2$, and perpendicular to the propagation vector. The Alfvén wave velocity only depends on the local magnetic field intensity and plasma mass density; thus there is no dispersion (frequency dependence) and the wave profile remains the same as it propagates. Historically, it was soon recognized that the so-called micropulsations of the ground-based geomagnetic field were standing oscillations of field-aligned Alfvén waves—making the magnetosphere a planetary-scale "musical instrument" of vibrating field lines. At the time of the discovery of these ultra low frequency (ULF) Alfvén waves, it was quite difficult to imagine the possibility of a wave propagation process in a collisionless gas—Alfvén waves became a prime example of the intricate interplay between currents and fields in a collisionless plasma. And, as we have mentioned in Chap. 4, they indeed play a fundamental role in the dynamics of the radiation belt.

In this chapter we just gave a somewhat superficial description aimed at showing how collisionless plasmas can be understood intuitively by focusing on the fundamental properties of the adiabatic behavior of charged particles in magnetic and electric fields. Formal and detailed descriptions can be found, for instance, in [2] (includes the most important relativistic equations), [3] and [4].

5.6 Epilogue

In his waning days Alfvén insistently lamented to one of us (JGR) about having promulgated the concept of "frozen-in magnetic field" and "moving field lines" too much during the early times of space plasma physics. He fully recognized that the field line is a purely geometric concept that can be very helpful in visualizing magnetic field geometry and, in *certain* situations, its time-changes, but that this image must be handled with great care. Field lines do not drag plasma, nor does plasma drag field lines—plasma moves in response to magnetic and electric forces acting on currents and charges embedded in the fluid, a process mathematically described by linking plasma and Maxwell's equations. It so happens that *under certain circumstances* we can visualize in our minds this motion as that of continuously changing magnetic field lines, co-moving with the plasma or, rather, its constituent electron fluid (see (5.53)).

As a tribute to Alfvén, let us end the book with the "grand finale" of a kindergarten example. Turn back to Fig. 5.2, and assume that now we have a *neutral* dense ensemble of 90° particles evenly distributed in that cylinder, in an external homogeneous magnetic field $\boldsymbol{B}$. To avoid undesirable equivalent polarization charges and other complications, we'll assume it to be a low-beta plasma ($p \ll B^2/(2\mu_0)$, page 136). The field inside the cylinder $\boldsymbol{B}^* = \boldsymbol{B} + \boldsymbol{b}$ will be reduced in intensity due to the diamagnetic effect of the boundary equivalent currents; the field topology of the self-field $\boldsymbol{b}$ is in effect that of a solenoid, opposed to $\boldsymbol{B}$ inside. This means that the total field will exhibit field lines bent somewhat outwards all along the lateral boundary surface, leaving a reduced flux inside. Now we turn on a uniform electrostatic field, say, perpendicular into the paper in Fig. 5.2, $\boldsymbol{E} \perp \boldsymbol{B}$. Obviously, the circling particles will all drift to the right with the same speed $V_E = E/B$, independent of their energy, mass and charge (Sect. 1.3). Will they carry with them the magnetic field lines? According to our probe particle definition of field line velocity (1.38), the answer is *no*! This definition indeed mandates (see page 20) that we turn off all contributions from potential electric fields, and examine the probe particle drift *exclusively* under the action of the $-\partial \boldsymbol{A}/\partial t$ induced electric field. And, carefully depicting in our mind the rigidly drifting axisymmetric $\boldsymbol{A}$-vector field configuration of the equivalent current system, regions of appreciable $-\partial \boldsymbol{A}/\partial t$ will only be found in the vicinities of this *moving* cylindrical surface current system. In other words, the plasma will drift to the right in the figure (a kindergarten version of plasma propulsion motor!) and open its way through the external $\boldsymbol{B}$-field lines as if you were walking though a corn field by bending the stocks around you. Field lines will never detach from the original magnet but just bend out and snap back as the cylindrical surface current system moves by; this applies to field lines both outside and inside the cylinder. *At no time will this plasma be carrying any frozen magnetic field lines with it*![14]

[14]All this is valid only for a low energy density, i.e., low-beta plasma. At higher densities the situation changes considerably. For instance, the equivalent current envelope may be intense enough so that the inner field $\boldsymbol{B}^*$ is so weak that the guiding center approximation breaks down and a kinetic description is necessary; in that case we can no longer talk about a common electric drift. Moreover, if the boundary current is intense enough, field line *loops* may appear enclosing parts of the equivalent current system and indeed move together with the bulk motion of the latter; this happens with the plasmoids in the magnetospheric tail or the solar magnetic loops that detach from photospheric loops to form the initial stage of a solar mass ejection. In summary, whether "magnetic field lines carry plasma" or "plasma carries magnetic field lines" depends entirely on the characteristics and the dynamic behavior of the *currents* around which those field lines are wound (remember that a magnetic field line is always part of a closed loop because of $\boldsymbol{\nabla} \cdot \boldsymbol{B} = 0$—even if that closure involves an infinite number of turns or occurs at infinity!).

References

1. V.M. Vasyliūnas, Electric field and plasma flow: what drives what? Geophys. Res. Lett. **28**, 2177–2180 (2001)
2. B. Rossi, S. Olbert, *Introduction to the Physics of Space* (McGraw Hill, New York, 1970)
3. G.L. Siscoe, Solar system magnetohydrodynamics, in *Solar-Terrestrial Physics: Principles and Theoretical Foundations*, ed. by R.L. Carovillano, J.M. Forbes. Volume 104 of Astrophysics and Space Science Library (Reidel, Dordrecht/Holland, 1983), pp. 11–100
4. D.R. Nicholson, *Introduction to Plasma Theory* (Wiley, New York, 1983)

Appendices

A.1 What You Should Know About *B* but Maybe Forgot

This book deals with the dynamics of electrically charged particles in the radiation belts and collisionless plasmas. We are interested in certain common physical characteristics of their motions in certain configurations of magnetic (and sometimes also electric) fields under certain a priori limiting conditions of energy and spatio-temporal scales. In most chapters the fields are *given*, i.e., controlled by sources external to the regions of interest, such as the Earth's internal core currents and the currents in the ionosphere, outer magnetosphere and interface with the solar wind. Only in the last chapter do we consider the contributions to the fields by the particles themselves. This appendix should serve *not* as a boring presentation of the algebraic, differential and geometric vector relationships between magnetic field quantities needed in the main text, but, rather, as a discussion of the *physical meaning and field-topological consequences* of such relationships. They are important not only from the mathematical point of view but should also be of interest to the experimentalist—indeed, of all space variables of interest, the magnetic field vector is perhaps the one most accessible to accurate, global and standardized measurements. In addition to these more "practical" aspects, in this appendix we will briefly elaborate on Maxwell's equations and some important conceptual aspects thereof that are germane to a better understanding of cause-and-effect relationships in magnetospheric plasma physics.

A.1.1 Magnetostatics in a Nutshell

Much of this book deals with cases in which a *static* magnetic field imposed from outside provides the organizing structure within which ensembles of individual particles move. For reasons that will become apparent further below, we shall introduce the equations ruling magneto- and electrostatics in the following, less

J.G. Roederer and H. Zhang, *Dynamics of Magnetically Trapped Particles*, Astrophysics and Space Science Library 403, DOI 10.1007/978-3-642-41530-2,

traditional way. We begin with an "ideal experiment" (by which we mean an experiment feasible in principle that would come out the way we claim, but which would be difficult to conduct accurately in practice). Let us assemble a given macroscopic distribution of charges and currents in a limited portion of 3-D space by bringing in charge elements $\delta q = \rho dr^3$ from infinity, and turning on closed current loops made of current elements $\delta \boldsymbol{j} = \boldsymbol{J} dr^3$ with some e.m.f. (e.g., batteries). We do this very, very slowly, holding the charges in equilibrium at all times by some external mechanical forces, and registering the mechanical or chemical energy delivered or extracted by the e.m.f.'s (assuming that no other matter is present, there should be no ohmic losses or polarization effects). Once the charges and current loops are assembled, we make sure, again through external forces, that the system remains in a steady equilibrium state. The total mechanical work U_e and U_m that must be done to assemble these distributions in vacuum turns out to be

$$U_e = \frac{1}{4\pi\epsilon_0} \frac{1}{2} \iint \frac{\rho(\boldsymbol{r})\rho(\boldsymbol{r}')}{|\boldsymbol{r} - \boldsymbol{r}'|} dr^3 dr'^3 \tag{A.1}$$

$$U_m = \frac{\mu_0}{4\pi} \frac{1}{2} \iint \frac{\boldsymbol{J}(\boldsymbol{r}) \cdot \boldsymbol{J}(\boldsymbol{r}')}{|\boldsymbol{r} - \boldsymbol{r}'|} dr^3 dr'^3 \tag{A.2}$$

These are sextuple integrals over the spatial coordinates $\boldsymbol{r}$ and $\boldsymbol{r}'$, representing the linear sum of *mutual, symmetric interactions* of pairs of static charge and current elements, respectively. The constants of proportionality $1/4\pi\epsilon_0$ and $\mu_0/4\pi$ universally represent the intensity of electrostatic and magnetostatic interactions. They are written to conform to traditional notation; each ϵ_0 and μ_0 must be separately determined experimentally once a unit of charge is adopted by international agreement (the unit of current density J follows from $\boldsymbol{J} = \rho \boldsymbol{v}$).[1] Notice in (A.1) and (A.2) that none of two interacting elements is "privileged" in any way over its partner—electro/magnetostatic interactions have no cause-effect direction, or preferred "source-probe" relationship. The singularity at $|\boldsymbol{r} - \boldsymbol{r}'| = 0$ is no problem as long as ρ and $\boldsymbol{J}$ are "well-behaved" continuous functions.

In reality, ρ and $\boldsymbol{J}$ are densities, i.e., *macroscopic* physical quantities; each element of volume dr^3, when examined with a magnifying glass, will be seen to contain huge amounts of microscopic "point" particles. In other words, when using the functions ρ and $\boldsymbol{J}$ in electromagnetism we are replacing "microscopic reality" (a statistical distribution of particles) with a *model*, a *virtual continuum* of electrically charged matter; ρ and $\boldsymbol{J}$ are local *space-time averages* of point charges in an element of volume dr^3. This is neither mathematical nor philosophical nitpicking—it is quite germane to understanding correctly the fundamental relations (A.1)

[1] In the SI system of units, $\epsilon_0 = 8.85 \times 10^{-12}\,(\mathrm{C})^2\,(\mathrm{N})^{-2}\,\mathrm{m}^{-2}$, $\mu_0 = 4\pi 10^{-7}\,(\mathrm{N})\,(\mathrm{C})^{-2}\,\mathrm{s}^2$ and the unit of charge (the Coulomb) is that of 6.2418×10^{18} protons (number of elementary charges which makes the figure in the value of μ_0 come out *exactly* $4\pi 10^{-7}$).

and (A.2): they represent the total energy of assembling macroscopic charge and/or current distributions, but they do *not* include the energy needed to create the microscopic charges themselves![2]

In (A.2) and (A.1) the charge and current densities are not completely independent: although we have assumed quasi-equilibrium, they must obey the most absolute conservation principle of physics, namely that the total electric charge in a closed system is constant, or, in general form:

$$\frac{d}{dt}\int_V \rho dr^3 = -\oint_\Sigma \boldsymbol{J}\cdot d\boldsymbol{S} \tag{A.3}$$

where Σ is a fixed, closed surface and V the volume enclosed. It is this conservation theorem that requires the distribution of currents to consist of *closed* current loops ($\nabla \boldsymbol{J} = 0$) in order to secure a stationary state. There is another important point to make. In our "ideal" experiment above we assumed that the current loops and charges were given, i.e., assembled by means of outside intervention—actions controlled by, and energy delivered by, systems *external* to the electromagnetic environment under consideration. In the real world, in presence of other matter, we also encounter macroscopic currents and charges which appear as a function of the local electromagnetic environment (field). Among them are the so-called *equivalent currents and charges* (the magnetization currents and polarization charges in certain materials and plasma); although they do not represent macroscopic transport or accumulation of individual charges, their macroscopic electromagnetic effects are similar to those of "real" charge *convection currents* ($\boldsymbol{J}_c = \rho \boldsymbol{v}$) and "free charge" density (ρ).[3] In what follows we shall ignore this distinction for the time being.

---

[2]The following example will illuminate this better. Consider two separate spheres uniformly charged (a macroscopic model!) with total charge $+Q$ and $-Q$, respectively. ρ at any interior point will be given by Q/[volume of sphere]. The total energy as given by the integral (A.1) can be divided into two big positive parts corresponding to the integrals over the pairs $\boldsymbol{r}, \boldsymbol{r}'$ that lie *within* each sphere (energy needed to assemble each sphere in isolation), and a much smaller third *negative* component containing the split pairs with $|\boldsymbol{r} - \boldsymbol{r}'|$ going from one point $\boldsymbol{r}$ in one sphere to another one $\boldsymbol{r}'$ in the other (energy needed to bring the already assembled charged spheres together). If one shrinks the size of the spheres to a point while maintaining the charge $|Q|$, those big positive contributions will run into infinity. Now, consider the fact that by using delta functions, a single point charge at $\boldsymbol{r}_0$ can be expressed as a charge *density* in the form $\rho(\boldsymbol{r}_0) = q\,\delta(\boldsymbol{r} - \boldsymbol{r}_0)$ (the defining property of the delta-function is $\int_{-\infty}^{+\infty} f(x)\delta(x-a)dx = f(a)$). Likewise, the current density of a moving point charge is $\boldsymbol{J}(\boldsymbol{r}_0) = q\boldsymbol{v}\,\delta(\boldsymbol{r} - \boldsymbol{r}_0)$. For an ensemble of point charges q_i, the integral in (A.1) will turn into sums over the point charges: $U_e \sim \Sigma(q_i q_k)/|\boldsymbol{r}_i - \boldsymbol{r}_k|$—but we must eliminate the singularities at $\boldsymbol{r} = \boldsymbol{r}'$ by imposing the condition $i \neq k$. This is like leaving out the self-energy of both spheres in the preceding example, with the total energy U_e coming out negative. Note that it is impossible to create a single charge "from scratch", but that we *can* create an element of current by just setting a charge in motion.

[3]Equivalent currents and charges can also be *external* sources of a magnetic or electric field: the magnets and electrets are materials of "given" quasi-permanent magnetization and polarization, respectively.

It is convenient to define the potentials V and $\boldsymbol{A}$ as scalar and vector functions of position:

$$V(\boldsymbol{r}) = \frac{1}{4\pi\epsilon_0}\int \frac{\rho(\boldsymbol{r}')}{\left|\boldsymbol{r}-\boldsymbol{r}'\right|}dr'^3 \tag{A.4}$$

$$\boldsymbol{A}(\boldsymbol{r}) = \frac{\mu_0}{4\pi}\int \frac{\boldsymbol{J}(\boldsymbol{r}')}{\left|\boldsymbol{r}-\boldsymbol{r}'\right|}dr'^3 \tag{A.5}$$

and express the total energies in simpler terms:

$$U_e = \frac{1}{2}\int \rho(\boldsymbol{r})V(\boldsymbol{r})dr^3 \quad \text{and} \quad U_m = \frac{1}{2}\int \boldsymbol{J}(\boldsymbol{r})\cdot\boldsymbol{A}(\boldsymbol{r})dr^3 \tag{A.6}$$

It can be deduced from the above energy equations (a tricky task for closed current distributions!) that the electro- and magnetostatic force densities at a given point of a charge and current distribution are, respectively,

$$\boldsymbol{f}_e = -\rho\nabla V \tag{A.7}$$

$$\boldsymbol{f}_m = \boldsymbol{J}\times(\nabla\times\boldsymbol{A}) \tag{A.8}$$

It also follows that the functions qV and $q\boldsymbol{v}\cdot\boldsymbol{A}$ appear in the Hamiltonian of mechanical systems with point charges—a fact that confers to the electromagnetic potentials a physical meaning far beyond that of merely being mathematically convenient functions.[4] Finally, we introduce the *electrostatic and magnetostatic fields*

$$\boldsymbol{E} = -\nabla V \quad \text{and} \quad \boldsymbol{B} = \nabla\times\boldsymbol{A}$$

so that the total electromagnetic force density acting on a distribution of charges and currents can be written as $\boldsymbol{f} = \rho\boldsymbol{E} + \boldsymbol{J}\times\boldsymbol{B}$, which for just one point charge leads to (1.1).[5] For the field vectors we can verify that the following first-order differential equations are a consequence of (A.1) and (A.2):

$$\nabla\cdot\boldsymbol{B} = 0 \tag{A.9}$$

$$\nabla\times\boldsymbol{B} = \mu_0\boldsymbol{J} \tag{A.10}$$

[4]Contrary to the scalar potential V, which plays a direct practical role because it can be manipulated in the lab, the vector potential $\boldsymbol{A}$ plays a lesser practical role in magnetostatics.

[5]The fields thus defined are *macroscopic* entities inside a charge and current distribution—i.e., they must be viewed as fields in a *virtual continuum* of charge and current. The actual "real" microscopic fields inside a material medium are a wild bunch, fluctuating around from zero to near-infinity as point charges fly by. Older $E\&M$ textbooks spend many pages with rigorous proofs that inside a medium the $\boldsymbol{E}$ and $\boldsymbol{B}$ vectors in Maxwell's equations are indeed space-time averages of the microscopic fields.

$$\nabla \times \boldsymbol{E} = 0 \tag{A.11}$$

$$\nabla \cdot \boldsymbol{E} = \frac{1}{\epsilon_0}\rho \tag{A.12}$$

These are the Maxwell's equations for *static* fields in vacuum. Turning around the issue (as it is normally presented in textbooks), given a distribution of charges ρ and currents $\boldsymbol{J}$ in space, the general solution of these differential equations are, in terms of the potentials, the integrals (A.4) and (A.5).

Concerning the current density $\boldsymbol{J}$ in (A.10), we mentioned that in presence of matter there are two distinct types: (i) *convection currents*, i.e., collectively moving charges $\boldsymbol{J}_c = \rho \boldsymbol{V}$, and (ii) *equivalent currents* $\boldsymbol{J}_e = \nabla \times \boldsymbol{M}$ due to matter of magnetization density $\boldsymbol{M}$ and/or an ensemble of charged particles in cyclotron motion (1.26). Although not an issue in this book, also the charge density ρ can be of two types, *free charges* ρ and *equivalent charges* $-\nabla \cdot \boldsymbol{P}$, where $\boldsymbol{P}$ is the electric polarization density in an ensemble of charged particles or a material medium. In Chap. 5 we deal with quasi-*neutral* ensembles of charged particles, in which the total charge density is always extremely small, i.e., where $|\rho^+| \cong |\rho^-|$; in that case, convection currents can only flow when $\boldsymbol{V}^+ \neq \boldsymbol{V}^-$. And in presence of collisions, we will encounter a third kind, the *conduction* related to the electric field $\boldsymbol{J} = \sigma \boldsymbol{E}$. The magnetization and polarization densities as well as the conductivity may all be functions of corresponding local macroscopic fields, hence they are not independent variables; in a plasma or conducting fluid, even convection currents (bulk charge flows) may be field-dependent variables under certain circumstances.

The field vectors $\boldsymbol{E}$ and $\boldsymbol{B}$ are usually viewed as the principal physical magnitudes of electromagnetism since Maxwell's times. Indeed, they are defined in whole space and can be considered intuitively as local "ambassadors"—even in vacuum—of all distant electromagnetic sources. In the post-Faraday era a transition indeed has taken place from the "action-at-a-distance" point of view to the picture of local field→charge and charge→field actions. And yet, we wish to emphasize that, *in the end*, what is wanted in electromagnetism is quantitative information about effects on electrically charged *matter*—regardless of how we depict in our minds, and what physical properties we assign to, the empty space between clumps of that charged matter.[6] Indeed, forces and energy are the only physical magnitudes that can be directly measured, and electric charges attached to physical matter the only ones that we can manipulate from outside through non-electromagnetic interactions. To create a field, to delete a field, to change a field, to measure a field, to probe its alleged mechanical properties, *inevitably* requires the presence of electrically charged matter—and this is even true in *quantum* electrodynamics!

[6] Maxwell himself cautioned that "his theory was to be regarded as a mere picture of nature, a mere analogy, which at the moment allowed one to summarize all the phenomena in the most comprehensive way" (quoted in [1], page 68).

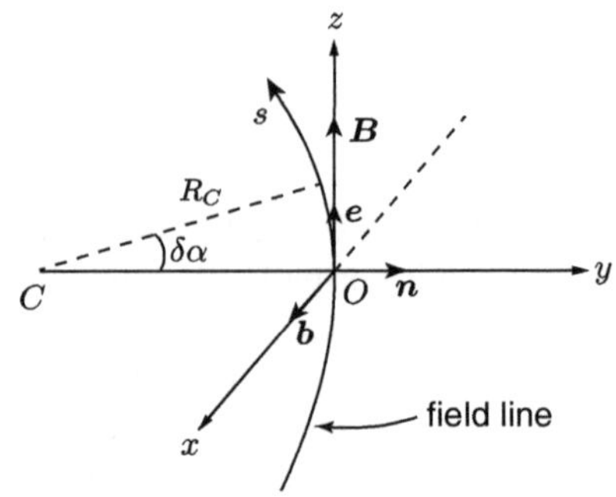

Fig. A.1 The local natural coordinate system

A.1.2 "Natural" Coordinate Systems in a Magnetic Field

In this section we shall examine some fundamental geometric/topological properties of a static magnetic field satisfying Eqs. (A.9), (A.10) and related relationships and physical meanings. First, we shall discuss the meaning of the spatial derivatives $\partial B_i/\partial x_k$. For this purpose we introduce a "natural coordinate system" as a local Cartesian system whose axes are oriented as shown in Fig. A.1. Clearly, there is a different system for each point of a $\boldsymbol{B}$–field; for this reason the natural system is only used as a coordinate system in the infinitesimal environment of the point in question (unless the field is homogeneous). The z-axis is parallel to $\boldsymbol{B}$ at O and the (y, z) plane is the *osculating plane* of the field line[7] through O (i.e., the limit of a plane defined by O and two field line points on either side, as the latter approach O). Axis x is normal to that plane. $\boldsymbol{e}$ is a unit vector parallel to $\boldsymbol{B}$; $\boldsymbol{n}$ lies in the osculating plane and is normal to the field line; the binormal $\boldsymbol{b}$ is perpendicular to both. We can write $\boldsymbol{e} = \boldsymbol{B}/B$; note that this expression is valid in *any* coordinate system. C is the center of curvature of the field line at O; δs is an element of arc length, and, by definition of radius of curvature R_c, we have $\delta s = R_c \delta\alpha$. The components of the magnetic field at the origin of a natural system are $B_x(0) = 0$, $B_y(0) = 0$ and $B_z(0) = B$. It is very important to realize that in a non-uniform magnetic field the natural system varies from point to point, i.e., that the unit vectors $\boldsymbol{b}, \boldsymbol{n}, \boldsymbol{e}$ have *non-zero spatial derivatives*.

[7]Field lines are geometric entities useful for the visualization of any vector field $\boldsymbol{F}(x, y, z)$. They are 3-D curves with the property that at each point the field vector is tangent: $\boldsymbol{F} = F\boldsymbol{e}$, where $\boldsymbol{e}$ is a unit vector tangent to the curve. A field line is defined through the geometric-differential relations $dx/F_x = dy/F_y = dz/F_z$ on each one of its points x_f, y_f, z_f. If we express the equations determining a 3-D curve as the pair of functions $x_f = x_f(z_f), y_f = y_f(z_f)$, the differential definition of a field line leads to the pair of simultaneous differential equations that define it: $dx/dz = F_x/F_z$ and $dy/dz = F_y/F_z$, with the initial condition given by the coordinates of the point through which the line should be traced. The right hand sides are known functions of x, y, z. The solutions x, y as a function of z are the equations of the field line.

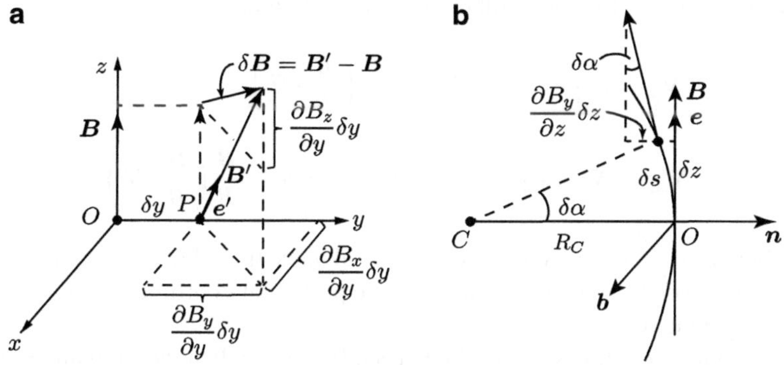

Fig. A.2 Depiction of geometric field and field line parameters

Figure A.2a, b show two aspects of the geometric meaning of the partial derivatives $\partial B_i/\partial x_k$ in the natural coordinate system. $\boldsymbol{B}'$ and $\boldsymbol{e}'$ are the field vector and the unit vector along it at a neighboring point P, on the y-axis in Fig. A.2a and along the field line in Fig. A.2b. There are some important "a priori" relationships (henceforth we shall use indistinctly δz or δs as the displacement along $\boldsymbol{B}$ or the field line):

$$\frac{\partial B_x}{\partial z} = 0 \quad \text{(by definition of osculating plane)} \tag{A.13}$$

$$\frac{\partial B_y}{\partial z} = -\frac{B}{R_c} \quad \text{(see Fig. A.2b)} \tag{A.14}$$

$$\frac{\partial \boldsymbol{e}}{\partial s} = -\frac{\boldsymbol{n}}{R_c} \quad \text{(by definition of curvature)} \tag{A.15}$$

From (A.9) we conclude that

$$\frac{\partial B_z}{\partial z} = \frac{\partial B}{\partial s} = -(\frac{\partial B_x}{\partial x} + \frac{\partial B_y}{\partial y}) \tag{A.16}$$

From all this we obtain the components of the curl of $\boldsymbol{B}$ in the natural system:

$$\nabla \times \boldsymbol{B} = \left(\frac{\partial B}{\partial y} + \frac{B}{R_c}\right)\boldsymbol{b} + \left(-\frac{\partial B}{\partial x}\right)\boldsymbol{n} + \left(\frac{\partial B_y}{\partial x} - \frac{\partial B_x}{\partial y}\right)\boldsymbol{e} \tag{A.17}$$

An often-used quantity is the gradient of the modulus B, or $\nabla B = [\partial B/\partial x, \partial B/\partial y, \partial B/\partial z]$. The first two components represent the vector $\nabla_\perp B = \partial B/\partial x\, \boldsymbol{b} + \partial B/\partial y\, \boldsymbol{n}$; the z component is $\nabla_\parallel B = \boldsymbol{e}\, \partial B/\partial s$. This allows us to introduce the gradient *operator* ∇ in its natural components:

$$\nabla = \nabla_\perp + \nabla_\parallel \tag{A.18}$$

which can be written for a general coordinate system as

$$\nabla_{\parallel} = e\frac{\partial}{\partial s} = e(e \cdot \nabla) = \frac{\boldsymbol{B}}{B}\left(\frac{\boldsymbol{B}}{B} \cdot \nabla\right) \tag{A.19}$$

$$\nabla_{\perp} = \nabla - \nabla_{\parallel} = \nabla - e(e \cdot \nabla) = \nabla - \frac{\boldsymbol{B}}{B}\left(\frac{\boldsymbol{B}}{B} \cdot \nabla\right) \tag{A.20}$$

In this business, it is indeed important to always find the most general vector expressions whose form is independent of the particular coordinate system in which they are formulated.

There is another expression for the perpendicular gradient valid in all coordinate systems, and used in several parts of the main text. It is obtained by crossing (A.17) with the unit vector $\boldsymbol{e}$, taking into account (A.15) and rearranging terms:

$$\nabla_{\perp} B = B\frac{\partial \boldsymbol{e}}{\partial s} - (\nabla \times \boldsymbol{B}) \times \boldsymbol{e} = (\boldsymbol{B} \cdot \nabla)\left(\frac{\boldsymbol{B}}{B}\right) - (\nabla \times \boldsymbol{B}) \times \left(\frac{\boldsymbol{B}}{B}\right) \tag{A.21}$$

In *absence of local currents* ($\nabla \times \boldsymbol{B} = 0$), we have

$$\nabla_{\perp} B = B\frac{\partial \boldsymbol{e}}{\partial s} = (\boldsymbol{B} \cdot \nabla)\left(\frac{\boldsymbol{B}}{B}\right) \quad \text{and} \quad \nabla_{\parallel} B = \boldsymbol{e}\frac{\partial B}{\partial s} = \left(\frac{\boldsymbol{B}}{B} \cdot \nabla\right)\boldsymbol{B} \tag{A.22}$$

The latter, because in view of (A.13) and (A.14), the components of $\partial \boldsymbol{B}/\partial s$ are $[0; -B/R_c; \partial B/\partial s]$. Another set of useful relations, trivially resulting from $\boldsymbol{B} = B\boldsymbol{e}$ is

$$\frac{\partial \boldsymbol{B}}{\partial s}|_{\perp} = \boldsymbol{e}\frac{\partial B}{\partial s} = \left(\frac{\boldsymbol{B}}{B} \cdot \nabla\right)\boldsymbol{B}$$

$$\frac{\partial \boldsymbol{B}}{\partial s}|_{\parallel} = B\frac{\partial \boldsymbol{e}}{\partial s} = (\boldsymbol{B} \cdot \nabla)\left(\frac{\boldsymbol{B}}{B}\right) \tag{A.23}$$

We now bring in more explicitly the current density $\boldsymbol{J}$. First we evaluate the curl of the unit vector $\boldsymbol{e}$. At first sight it may seem strange that the spatial derivative of a unit vector which determines the direction of an axis of a coordinate system is not zero. But remember that the coordinate system in question is *local*, i.e., changes from point to point—and with it, its axes change (see Fig. A.2). We have, taking into account that $\boldsymbol{B} \times \nabla \boldsymbol{B} = \boldsymbol{B} \times \nabla_{\perp} \boldsymbol{B}$ and making use of (A.21):

$$\nabla \times \boldsymbol{e} = \nabla \times \frac{\boldsymbol{B}}{B} = \boldsymbol{e} \times \frac{\partial \boldsymbol{e}}{\partial s} + \boldsymbol{e}\,\frac{\boldsymbol{B} \cdot (\nabla \times \boldsymbol{B})}{B^2}$$

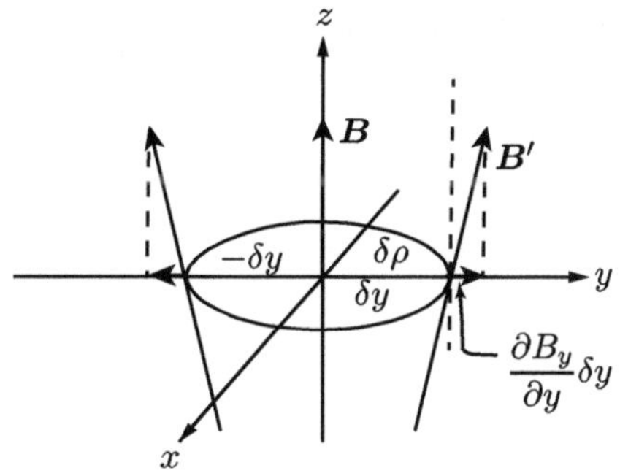

Fig. A.3 Divergence/convergence of field lines

This leads us to the pair of expressions:

$$\nabla \times \boldsymbol{e}\big|_{\perp} = \boldsymbol{e} \times \frac{\partial \boldsymbol{e}}{\partial s} \tag{A.24}$$

$$\nabla \times \boldsymbol{e}\big|_{\parallel} = \boldsymbol{e}\,\frac{\boldsymbol{e} \cdot (\nabla \times \boldsymbol{B})}{B} \tag{A.25}$$

In the main text, we deal frequently with the perpendicular and parallel components of the current density. From (A.21) and taking into account (A.20) and (A.25) for a static field, we obtain the pair:

$$\mu_0 \boldsymbol{J}_{\perp} = \nabla \times \boldsymbol{B}\big|_{\perp} = \boldsymbol{B} \times \left(\frac{\partial \boldsymbol{e}}{\partial s} - \frac{\nabla_{\perp} B}{B}\right) \tag{A.26}$$

$$= \frac{\boldsymbol{B}}{B} \times \left[\left(\frac{\boldsymbol{B}}{B} \cdot \nabla\right) B - \nabla B + (\boldsymbol{B} \cdot \nabla)\left(\frac{\boldsymbol{B}}{B}\right)\right]$$

$$\mu_0 \boldsymbol{J}_{\parallel} = \nabla \times \boldsymbol{B}\big|_{\parallel} = \boldsymbol{B}\big[(\nabla \times \boldsymbol{e}) \cdot \boldsymbol{e}\big] = B\left(\nabla \times \frac{\boldsymbol{B}}{B}\right) \tag{A.27}$$

Notice that the expression of $\boldsymbol{J}_{\parallel}$ cannot be reduced any further than just being the parallel component of the curl of $\boldsymbol{B}$. As shown in the main text, this is one of the fundamental properties responsible for the quite different character of field-aligned currents as compared to that of perpendicular currents in a plasma.

We now turn specifically to the geometric properties of a magnetostatic field, in particular the topology of its field lines, as determined by the spatial derivatives of its components. In Fig. A.3 we illustrate how the diagonal components $\partial B_i/\partial x_i$ control the *divergence or convergence* of magnetic field lines. $1/B\,(\partial B_x/\partial x)$ and $1/B\,(\partial B_y/\partial y)$ give the angular measure of this spread, and $1/B\,(\partial B_z/\partial z) = 1/B\,(\partial B/\partial s)$ that of the relative rate of change of the magnitude of $\boldsymbol{B}$ which assures the constancy of the flux through an infinitesimally thin expanding or constricting flux tube. Of course, because of (A.16), this trio of derivatives is not independent.

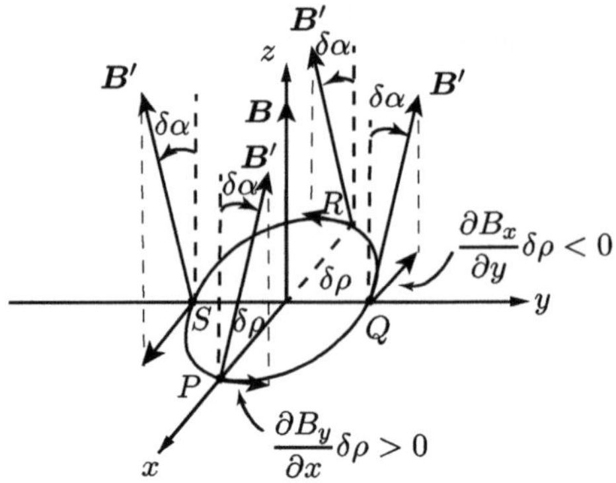

Fig. A.4 Torsion of field lines

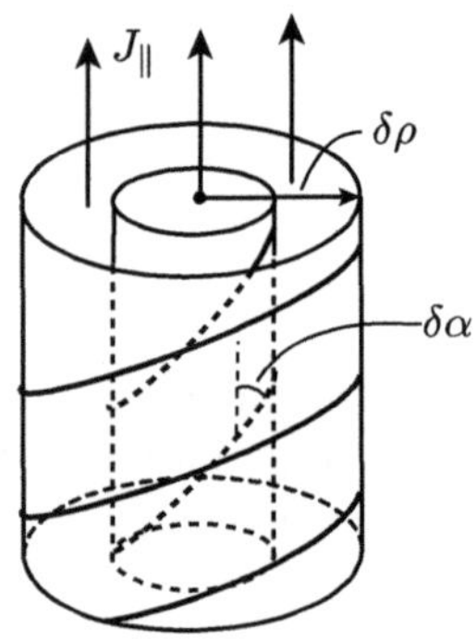

Fig. A.5 Torsion effect of a field-aligned current density tube

Next we consider the derivatives that appear in the parallel (z) component of the curl of $\boldsymbol{B}$: $\partial B_y/\partial x$ and $\partial B_x/\partial y$ (assume that all other derivatives are zero). Figure A.4 shows the case when both add up to zero:

$$\frac{\partial B_y}{\partial x} + \frac{\partial B_x}{\partial y} = 0$$

Observing the figure, this obviously means that the field lines surrounding the axis z are wound up, with a *torsion parameter* (absolute values only!)

$$\Theta = \frac{\delta\alpha}{\delta\rho} = \frac{1}{B}\frac{\partial B_y}{\partial x} = \frac{1}{B}\frac{\partial B_x}{\partial y} \tag{A.28}$$

In this case we must have a parallel current

$$\mu_0 J_{\parallel} = \frac{\partial B_y}{\partial x} - \frac{\partial B_x}{\partial y} = 2B\Theta \tag{A.29}$$

Do not think that this current only flows along the z-axis: it is a current *density*, and as we move away from that axis ($\delta\rho$ increases) the torsion increases and the *local* z or $\boldsymbol{n}$ axis gets more and more tilted (see sketch in Fig. A.5). So the picture becomes

Fig. A.6 Shear of field lines

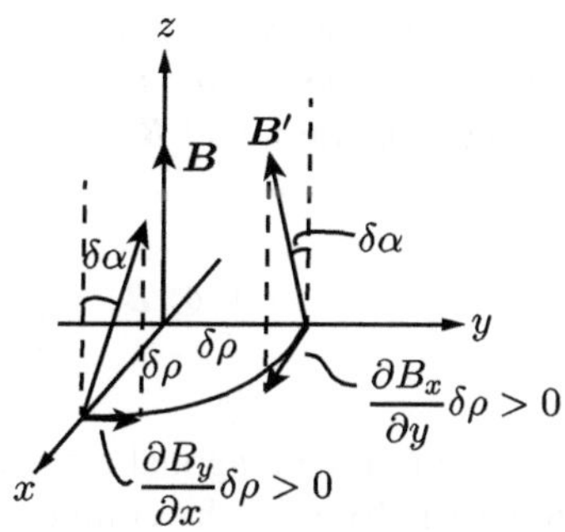

increasingly complicated and will depend on exactly how the derivatives in (A.28) themselves vary in space. The resulting *flux ropes* are topologically complicated entities, and applying intuition is indeed a dangerous game! The bottom line of all this, in particular relation (A.29), is that any field-aligned current $\boldsymbol{J}_{\parallel}$ causes torsion of neighboring magnetic field lines. The reverse, however, is not necessarily true: there exist field-line topologies that *look* like torsion but wherein $\nabla \times \boldsymbol{B} = 0$—e.g., the vacuum field around a current-carrying wire with a weak magnetic field parallel to the wire, or the vacuum field in a solenoid with a weak longitudinal linear current, as in some plasma fusion machines.

The other extreme case with these two derivatives arises when

$$\frac{\partial B_x}{\partial y} - \frac{\partial B_y}{\partial x} = 0$$

Now certainly there are no field-aligned currents; the effect on the field topology is that of *shear*—see Fig. A.6. The shear rate is defined as (absolute values only!)

$$\Sigma = \frac{\delta\alpha}{\delta\rho} = \frac{1}{B}\frac{\partial B_x}{\partial y} = \frac{1}{B}\frac{\partial B_y}{\partial x} \tag{A.30}$$

In this case

$$\frac{\partial B_y}{\partial x} + \frac{\partial B_x}{\partial y} = 2B\Sigma \tag{A.31}$$

We can summarize the torsion-shear cases in the relations:

$$\frac{\partial B_x}{\partial y} = B(\Sigma + \Theta) \tag{A.32}$$

$$\frac{\partial B_y}{\partial x} = B(\Sigma - \Theta) \tag{A.33}$$

It is instructive to pull together all the spatial derivatives of the magnetic field vector into the *gradient tensor* of the vector $\boldsymbol{B}$, which is the tensor product of the gradient operator and $\boldsymbol{B}$[8]:

$$\nabla \otimes \boldsymbol{B} = \begin{pmatrix} \partial B_x/\partial x & \partial B_y/\partial x & \partial B_z/\partial x \\ \partial B_x/\partial y & \partial B_y/\partial y & \partial B_z/\partial y \\ \partial B_x/\partial z & \partial B_y/\partial z & \partial B_z/\partial z \end{pmatrix} \tag{A.34}$$

The sum of the diagonal elements or *trace* of this tensor, a scalar, is zero according to (A.9). To write the components of this tensor in the *natural* coordinate system, we can take into account relations (A.14), (A.16), (A.32) and (A.33):

$$\nabla \otimes \boldsymbol{B} = \begin{pmatrix} \partial B_x/\partial x & B(\Sigma - \Theta) & \nabla_{\perp x} B \\ B(\Sigma + \Theta) & -(\partial B/\partial s + \partial B_x/\partial x) & \nabla_{\perp y} B \\ 0 & -B/R_c & \partial B/\partial s \end{pmatrix} \tag{A.35}$$

In absence of local currents, Θ and $\nabla_{\perp x} B$ are both zero, $\nabla_{\perp y} B = -B/R_c$ and the gradient tensor is symmetric. The condition for a uniform magnetic field is $\nabla \otimes \boldsymbol{B} \equiv 0$ in a finite region of space.

In Chap. 5 we will come across an expression that involves the divergence operator acting on a second order tensor $\mathbb{T}$, of components $\nabla \mathbb{T}|_i = \Sigma_k \partial/\partial x_k T_{ik}$. Using (A.26) and (A.21) we can verify that

$$\mu_0 \boldsymbol{J} \times \boldsymbol{B} = \mu_0 \boldsymbol{J}_\perp \times \boldsymbol{B} = -\nabla(\frac{B^2}{2}) + (\boldsymbol{B} \cdot \nabla)\boldsymbol{B} \tag{A.36}$$

The components of the first term can be written

$$\frac{\partial}{\partial x_i}(\frac{B^2}{2}) = \Sigma_k \frac{\partial}{\partial x_k}\left(\frac{B^2}{2}\delta_{ik}\right)$$

[8] A *tensor* of second rank $\mathbb{T}$ is a mathematical entity of nine components T_{ik} which linearly assigns to any vector $\boldsymbol{P}$ another vector $\boldsymbol{Q}$ in the form $Q_i = \sum_k T_{ik} P_k$ (in Cartesian space). To assure that the correspondence $P \rightarrow Q$ is independent of the particular coordinate system when changing from the original Cartesian system O to a rotated system O*, the components of the tensor must be transformed according to the rule $T^*_{mn} = \sum_{ik} \gamma_{mi}\gamma_{nk} T_{ik}$, where the γ's are the cosines of the angles between the coordinate systems' axes: $\gamma_{rs} = \boldsymbol{e}^*_r \cdot \boldsymbol{e}_s$. The *tensor product* between two vectors $\boldsymbol{A} \otimes \boldsymbol{B}$ is a tensor whose components are $T_{ik} = A_i B_k$. The product of a tensor with a vector (also called contraction) is a vector $\boldsymbol{B} = \mathbb{T}\boldsymbol{A}$ with components $B_i = \sum_k T_{ik} A_k$ (the very definition of a tensor). In a Cartesian system there coexists a second version of the product, with components ${B_k}^T = \sum_i T_{ik} A_i$ (sum over the *first* index of the tensor); for a symmetric tensor, of course, $\boldsymbol{B} = \boldsymbol{B}^T$. For the same reason, there are *two* gradients of a tensor: the one shown in (A.34) and one transposed, of components $\nabla \otimes \boldsymbol{B}|^T_{ik} = \partial B_i/\partial x_k$.

where $\delta_{ik} = 1$ if $i = k$ and zero otherwise. Likewise, taking into account (A.9) we can verify that

$$(\boldsymbol{B} \cdot \nabla)\boldsymbol{B}|_i = \Sigma_k \frac{\partial}{\partial x_k}(B_i B_k)$$

The preceding two expressions lead us to an important relation for the magnetic force density f_m, which in components reads

$$f_i = \boldsymbol{J} \times \boldsymbol{B}|_i = \Sigma_k \frac{\partial}{\partial x_k} \frac{1}{\mu_0} \left[B_i B_k - \frac{B^2}{2} \delta_{ik} \right] \tag{A.37}$$

The quantities

$$\mathbb{S}_{m,ik} = \frac{1}{\mu_0} \left[B_i B_k - \frac{B^2}{2} \delta_{ik} \right] \tag{A.38}$$

are components of Maxwell's magnetostatic field *stress tensor*.[9] So in vector form:

$$\boldsymbol{f}_m = \boldsymbol{J} \times \boldsymbol{B} = \nabla \mathbb{S}_m \tag{A.39}$$

where

$$\mathbb{S} = \frac{1}{\mu_0} \boldsymbol{B} \otimes \boldsymbol{B} - \frac{B^2}{2\mu_0} \mathbb{I}$$

Finally, it can be shown that relation (A.2) can be converted into an integral over all space of the quantity $u = (1/2\mu_0) B^2$:

$$U_m = \int_{-\infty}^{+\infty} \frac{B^2}{2\mu_0} dr^3 \tag{A.40}$$

This is why u is called magnetic energy density. Notice that (A.38) and (A.40) are defined in *all* of space, i.e., also in vacuum. This is the basis of Maxwell's fundamental contribution to the "mechanification" of the electromagnetic field, assigning it properties similar to those of an elastic medium (but beware: this is just a *model* ... !).

---

[9]There is another equivalent stress tensor for the electric field: $\mathbb{S}_{e,ik} = \epsilon_0 [E_i E_k - (E^2/2)\,\delta_{ik}]$.

A.1.3 Electrodynamics in a Nutshell: Interpreting Maxwell's Equations

Let return to the relations (A.1) and (A.2). We introduced them as expressions of the energy (work of external forces and batteries) needed to build (very slowly, always in quasi-equilibrium) *stationary* distributions of charges ρ and currents $\boldsymbol{J}$. We stated that from the physical point of view, these expressions represent the mutual electrostatic and magnetostatic interactions between elements of charge (at rest or moving in closed current loops). What if we lifted the restriction of slow, adiabatic assembly, and the distributions are no longer stationary in the end state? We must turn to "ideal experiments" again, which will tell us that the fundamental expressions (A.4) and (A.5) are still valid, but need a small yet very significant correction:

$$V(\boldsymbol{r},t) = \frac{1}{4\pi\epsilon_0}\int \frac{\rho(\boldsymbol{r}',t_{ret})}{|\boldsymbol{r}-\boldsymbol{r}'|}dr'^3 \tag{A.41}$$

$$\boldsymbol{A}(\boldsymbol{r},t) = \frac{\mu_0}{4\pi}\int \frac{\boldsymbol{J}(\boldsymbol{r}',t_{ret})}{|\boldsymbol{r}-\boldsymbol{r}'|}dr'^3 \tag{A.42}$$

in which

$$t_{ret} = t - \frac{|\boldsymbol{r}'-\boldsymbol{r}|}{c}$$

is the *retarded time*, i.e., the time retarded by what it takes electromagnetic information to travel with speed c from point $\boldsymbol{r}'$ to $\boldsymbol{r}$. In other words, electromagnetic interactions between charges and currents are not instantaneous: an electric charge or current is not affected by what is happening elsewhere *now*, but what happened at other places *earlier*—by the time it takes electromagnetic information to travel from those other places to the point in question. This is no different than what happens with the acoustical sound field at any given time t: you hear what has been emitted from any other point in space not at time t (as it is always wrongly shown in the movies!) but at a *retarded* time ($c_{sound} \simeq 330\,\text{m/s}$).

The law of conservation of electric charge (A.3) remains as stated, but we must emphasize that in this latter expression *everything* is taken at the simultaneous present time t.[10] A fundamental fact is that this conservation law *links* the two

[10] This is not a trivial statement, especially having the theory of relativity in mind, which tells us that simultaneity depends on the frame of reference. For instance, take a distribution of charges $\rho(\boldsymbol{r})$ moving rigidly with a common velocity $\boldsymbol{V}$ to the right. Seen from the original frame of reference (OFR), there is as a convective current distribution $\boldsymbol{J} = \rho\boldsymbol{V}$ between the left and right edges of the charge distribution. At the right extreme, again as viewed by an OFR observer, charges will gradually appear in initially uncharged elements of volume fixed to the OFR; at the left extreme, fixed volume elements initially covered by the distribution will loose electric charges ($\partial\rho/\partial t \neq 0$ there!). The corresponding balance between current and time-rate of change of charge must be taken *at the same instant of time* left and right, *in the OFR*!

potentials. The distribution of currents in (A.42) can now consist of *open* current loops—e.g., current filaments running from one charge accumulation to another, provided (A.3) is satisfied. All this leads to a surprising relation between the velocity c and the "static" universal constants ϵ_0 and μ_0. Indeed, the following are *consequences* of (A.41), (A.42) and (A.3)[11]:

$$\nabla \cdot \boldsymbol{A} + \frac{1}{c^2}\frac{\partial V}{\partial t} = 0 \quad \text{(called Lorentz gauge)} \tag{A.43}$$

$$c = \frac{1}{\sqrt{\epsilon_0 \mu_0}} \quad \text{(a universal constant)} \tag{A.44}$$

Another result is that the electrical force density (A.7) now becomes

$$\boldsymbol{f}_e = -\rho\big(\nabla V + \frac{\partial \boldsymbol{A}}{\partial t}\big) \tag{A.45}$$

The additional term is an expression of Faraday's induction law: an electric charge at rest is subjected to an additional force in a changing magnetic field. For the same reason, the electric field $\boldsymbol{E} = \boldsymbol{f}_e/\rho$ must be redefined, while the expression for the magnetic field remains the same:

$$\boldsymbol{E} = -\nabla V - \frac{\partial \boldsymbol{A}}{\partial t} \tag{A.46}$$

$$\boldsymbol{B} = \nabla \times \boldsymbol{A} \tag{A.47}$$

The two terms in the electric field can be called the potential and induced parts, respectively. The first one originates in all electric charges, the second one in all changes of currents.[12]

With the above definitions and relationships, Maxwell's equations in vacuum follow at once. There are two ways of writing them, each way conveying a specific

---

[11]Just apply the operators $\partial/\partial t$ and $\nabla\cdot$ to the integrals (A.41) and (A.42), respectively, and then use the differential form of (A.3) (once inside the integral, switch the operator ∇ to one with respect to the variable $\boldsymbol{r}'$: $\nabla' = -\nabla$) [2].

[12]Note that in the approach followed above, with the potentials defined a priori in (A.41) and (A.42), the time derivative $\partial\lambda/dt$ and gradient $-\nabla\lambda$ (with λ an arbitrary scalar function) that respectively can be added to V and $\boldsymbol{A}$ so as to yield the *same* fields, are *zero* (Lorentz gauge (A.43)). Still, physically it is not possible to separate the two parts of $\boldsymbol{E}$ in (A.46) unless one makes the comparison with a system in which all electric charges have been eliminated, leaving only a time-dependent current system. The induced part of the electric field is crucial, however, for the definition of magnetic field line motion, as discussed in Chaps. 1 and 5.

meaning. The usual presentation[13] is the following (as always using rationalized SI units):

$$\nabla \cdot \boldsymbol{B} = 0 \tag{A.48}$$

$$\nabla \times \boldsymbol{B} = \mu_0 \boldsymbol{J} + \mu_0 \epsilon_0 \frac{\partial \boldsymbol{E}}{\partial t} \tag{A.49}$$

$$\nabla \cdot \boldsymbol{E} = \frac{1}{\epsilon_0} \rho \tag{A.50}$$

$$\nabla \times \boldsymbol{E} = -\frac{\partial \boldsymbol{B}}{\partial t} \tag{A.51}$$

The first Eq. (A.48) can be considered a direct consequence of the fourth, so only three of the above equations are mathematically independent. The differential form of the conservation of electric charge

$$\nabla \cdot \boldsymbol{J} = -\frac{\partial \rho}{\partial t} \tag{A.52}$$

can be added as a dependent fifth equation.

Equations (A.48) and (A.50) above tell us, respectively, that magnetic monopoles (the magnetic equivalents of electric charges) do not exist and that the sources of the electric field are charge distributions of density ρ. Therefore, the magnetic field is purely rotational, with vorticity originating in the electric currents (of density $\boldsymbol{J}$) and in the time-changes of the electric field (the "displacement current"), according to the second Eq. (A.49). The fourth Eq. (A.51) tells us that in a stationary case the electric field is conservative, but that in a time-dependent situation it has a vorticity linked to the changing magnetic field (Faraday's law).

We may be tempted to interpret Maxwell's equations as showing that the electromagnetic field at any point $\boldsymbol{r}, t$ in space-time is determined by *local properties* like electric charge and current densities. This is not legitimate, however: what these equations really tell us is how the spatial and temporal *variations* of the fields are locally interconnected and affected by local electric charges and currents. Indeed, to find out the *actual* values of $\boldsymbol{E}$ and $\boldsymbol{B}$ vectors at any point $\boldsymbol{r}, t$ in space-time, one has to *integrate* the equations (via the electromagnetic potentials V and $\boldsymbol{A}$; see above). What ultimately counts in this operation are the charges $\rho(\boldsymbol{r}', t')$ and

[13]Ludwig Boltzmann once wrote [3]: "*War es ein Gott der diese Zeichen schrieb?*" (Was it a God who wrote these signs? ...). Indeed, not only does all of classical electromagnetism follow from these partial differential vector equations, but they also serve as the foundation of special relativity which revolutionized classical mechanics and the entire conception of space and time (tragically, Boltzmann hanged himself before he could find this out!)

currents $J(\boldsymbol{r}', t')$ *everywhere*, near and far,[14] taken at the appropriately retarded times $t' = t_{ret}$ as shown in (A.41) and (A.42).

The other way of writing down Maxwell's equations is by just rearranging the terms and presenting them as "evolution equations" (like the Hamilton and Schroedinger equations):

$$\frac{\partial \boldsymbol{B}}{\partial t} = -\nabla \times \boldsymbol{E} \tag{A.53}$$

$$\frac{\partial \boldsymbol{E}}{\partial t} = \frac{1}{\mu_0 \epsilon_0}(\nabla \times \boldsymbol{B} - \mu_0 \boldsymbol{J}) \tag{A.54}$$

$$\frac{\partial \rho}{\partial t} = -\nabla \cdot \boldsymbol{J} \tag{A.55}$$

with the following source equations as consequence of the above:

$$\nabla \cdot \boldsymbol{B} = 0 \qquad \nabla \cdot \boldsymbol{E} = \frac{1}{\epsilon_0}\rho$$

Presented in this fashion we can now legitimately state that what is determined by the local conditions are the local *time evolutions* of $\boldsymbol{E}$, $\boldsymbol{B}$ and ρ. For the magnetic field, the time evolution is governed by the *local vorticity* of $\boldsymbol{E}$ (which disappears in a static case); for the electric field, it is the *local imbalance* between the curl of $\boldsymbol{B}$ and the local current density (this imbalance is zero in a static case); and for the charge density it is the local net generation (or sink) of currents. However, we are still kicking the can down the road: vorticity and imbalance are still controlled by distant causes! And we must always remember that in a plasma (and any other material medium), equivalent currents appear that in themselves are functions of the local field that we are trying to determine, but which is a function of *all* distant sources. In Chap. 5 we shall address more explicitly this often misunderstood "chicken-and-egg" or cause-and-effect issue [4]: is it changes in the *currents* that lead to changes of the magnetic field or vice versa? Is it changes in the *electric field* that drive changes in currents and flows, or do changes in currents and flows shape the electric field?[15] The above evolutionary presentation of Maxwell's equations

[14]The case of a plane electromagnetic wave running from $-\infty$ to $+\infty$ (in which there are no sources anywhere because it is the solution of a homogeneous wave equation) is useful for teaching purposes but without any real physical meaning: there are no electromagnetic waves in nature *without* a source somewhere at some earlier time!

[15]An equivalent example of this question can be found in familiar electrostatics. Conductors are materials which do not admit spatial charge densities when in an equilibrium state—the free electrons distribute themselves on the surface so as to always maintain a zero electric field inside (i.e., constant electrostatic potential). Now take a system of several charged conductors in electrostatic equilibrium. The electric field in the vacuum between them may be extremely complex, if the shape of the surfaces is very complex. When you change the position (or charge) of any conductor, both the field and the surface densities on each conductor will change. Question:

will be helpful, but the inclusion of equivalent currents which do depend on the local field will be playing the deciding role.

Finally, a few remarks about the often orphaned magnetic vector potential. For a uniform magnetic field $\boldsymbol{B}_0$, the expression is

$$\boldsymbol{A}_0 = \boldsymbol{C} \times \boldsymbol{r} \tag{A.56}$$

where the vector $\boldsymbol{C}$ is a constant. To this expression the gradient of an arbitrary function can be added without changing the resulting value of $\boldsymbol{B}_0 = \nabla \times \boldsymbol{A}$; this will be necessary in non-symmetrical situations, in order to conform to the geometry of the current system giving rise to the uniform field (all uniform fields have a limit in the real world!). This is one of the tricky aspects of the vector potential: for instance, in (A.56) what is $\boldsymbol{r}$? For the field in a circular long solenoid it would be the radius vector from the axis (see Fig. 1.14); in that case $\boldsymbol{C} = \boldsymbol{B}_0$ (careful: $\nabla \cdot \boldsymbol{r} = 2$ because it is a *two*-dimensional radius vector!). For the near-uniform field close to a planar laminar 2-D current flow $\boldsymbol{r}$ would be the local vector distance to the flow surface, and $\boldsymbol{C} = \boldsymbol{B}_0$. The key for a uniform field is the linearity of the vector potential function $\boldsymbol{A}(\boldsymbol{r})$. Notice that (A.5) shows that, in general, the vector potential will tend to be parallel to the average of the current density distribution nearest in space and retarded time; the magnetic field, being the curl of $\boldsymbol{A}$, will tend to be perpendicular to $\boldsymbol{A}$; and an induced electric field will tend to be *perpendicular* to $\boldsymbol{B}$ if the variation of $\boldsymbol{A}$ is mainly in magnitude (the usual situation), and *parallel* to $\boldsymbol{B}$ if $\boldsymbol{A}$ only changes in direction (as in rotating magnetic field systems—e.g., see Fig. 2.3).

If the sources of a magnetic field are constant in time but move rigidly with constant velocity $\boldsymbol{V}_0$, the $\boldsymbol{A}$-vector in the original frame of reference (OFR) will vary locally only due to displacement:

$$\partial \boldsymbol{A}/\partial t = -(\boldsymbol{V}_0 \cdot \nabla)\boldsymbol{A} = \boldsymbol{V}_0 \times (\nabla \times \boldsymbol{A}) = \boldsymbol{V}_0 \times \boldsymbol{B} \tag{A.57}$$

For the second equality we took into account that in this case $\nabla \cdot \boldsymbol{A} = 0$. Since $\boldsymbol{E}_{ind} = -\partial \boldsymbol{A}/\partial t$, relation (A.57) leads precisely to the system transformation for the electric field given in (1.9). If now instead of moving, a uniform magnetic field is changing in magnitude $\dot{B} \neq 0$, the local induced electric field will be

$$\boldsymbol{E}_{ind} = -\dot{B}\ \boldsymbol{e} \times \boldsymbol{r} \tag{A.58}$$

---

does the change in electric charge density drive the change in the electric field, or is it the other way round? The answer is: Yes! Both are true! While it is unquestionable that the scalar potential of the electric field everywhere is *determined* (i.e., caused) by the electric charges as prescribed by (A.4), these charges, if subjected to a field inside the conductor will flow with a density $\boldsymbol{J} = \sigma \boldsymbol{E}_{inside}$ (σ is the conductivity) *until* they are all assembled on the surface in such a way as to zero out the field inside the conductor. In other words "changes in charge density drive changes in $\boldsymbol{E}$, changes in $\boldsymbol{E}$ drive changes in charge density, these changes in charge density drive changes in $\boldsymbol{E}, \ldots$"—until equilibrium is reached (a dynamic process that for good conductors takes fractions of a second).

A.1.4 The Mess with Electromagnetic Units: Why?

Students often complain about the "mess" with electromagnetic units. The reasons for the existence of several competing systems are found in the history of magnetism and electricity. It all boils down to the fact that they initially developed as independent branches of science. The following is a summary of the principal definitions and relations, translated into current terms and experiments.

Magnets have fascinated scientists for centuries with their capability of action at a distance (no apparent medium to transmit force and energy from one magnet to another). The concept of a magnetic field filling the empty space between magnets existed for a long time—especially when it was recognized that the Earth itself was a giant magnet. When laboratory experiments and methodical measurements became a fundamental ingredient of the scientific method, it was possible to define both, the magnetic dipole moment $\boldsymbol{m}$ and the magnetic field $\boldsymbol{H}$ through the relation $\boldsymbol{T}_m = \boldsymbol{m} \times \boldsymbol{H}$, where $\boldsymbol{T}_m$ is the torque on a little "probe" magnet in presence of a "source" magnet. Another set of experiments showed that the magnetic field at point $\boldsymbol{r}$ surrounding the source magnet depended on its magnetic moment $\boldsymbol{m}'$ and the position in the following way $\boldsymbol{H}(\boldsymbol{r}) = -\nabla(\boldsymbol{m}' \cdot \boldsymbol{r}/\mu^* r^3)$, with μ^* a universal constant (our notation! In reality the concept of "magnetic mass" preceded that of dipole moment).

Later came the experiments with electrically charged bodies, which if translated into experiments with electric *dipoles* of moment $\boldsymbol{p}$, could be summarized as $\boldsymbol{T}_e = \boldsymbol{p} \times \boldsymbol{E}$, thus defining the electric dipole moment and the electric field. The field of a source dipole $\boldsymbol{p}'$ turns out $\boldsymbol{E}(\boldsymbol{r}) = -\nabla(\boldsymbol{p}' \cdot \boldsymbol{r}/\epsilon^* r^3)$, with ϵ^* another universal constant (of course, historically, it was experiments with singly-charged bodies, leading to "Coulomb's law"). Then came "Oersted's surprise", namely that electric currents generate a magnetic field: a small electric current loop (of intensity I and area $\boldsymbol{a}$) behaves like a little magnet of magnetic moment $\boldsymbol{m} = \gamma^* I \boldsymbol{a}$, with γ^* yet another universal constant (again, historically the experiments involved linear conductors and the magnetic dipole unit was chosen so as to assure that $m = Ia$ without any new constant). The magnetic field of small current loops, however, had to be rotational (there was nothing physically at $r = 0$ to justify a singularity!), and so another vector was introduced, with the name of "magnetic induction" $\boldsymbol{B} = \mu^* \boldsymbol{H}$, which for a small source loop is given by the *divergence-free* relation $\boldsymbol{B} = \nabla \times (\gamma^* I \boldsymbol{a} \times \boldsymbol{r}/r^3)$.[16] On the other hand, the torque on a little current loop is $\boldsymbol{T}_m = \gamma^*/\mu^* \; I \boldsymbol{a} \times \boldsymbol{B}$. Finally, in our notation, the force densities in a distribution of charges and currents are $\boldsymbol{f}_e = \rho \boldsymbol{E}$ and $\boldsymbol{f}_m = \gamma^*/\mu^* \; \boldsymbol{J} \times \boldsymbol{B}$, respectively, and the electric force on a point charge is $\boldsymbol{F}_e = -q\nabla V$, the magnetic (Lorentz) force $\boldsymbol{F}_m = \gamma^*/\mu^* \; q\boldsymbol{v} \times \boldsymbol{B}$. Notice the following rule with our notation: (i) whenever a charge is *acted upon* by an electric field, it appears as q in the equations; whenever

[16]The above vector expressions for $\boldsymbol{H}$ (as a gradient) and $\boldsymbol{B}$ (as a curl) yield identical values except at the origin, where they differ by $-\boldsymbol{m}\,\delta(\boldsymbol{r})$ (which physically represents the singularity of the conservative field of a *point* dipole of magnetic "masses").

it is the *source* of an electric field it appears as q/ϵ^*; (ii) whenever a moving charge is *acted upon* by a magnetic field, it appears as $\gamma^*/\mu^* q\boldsymbol{v}$; when it is the *source* of a magnetic field, it is $\gamma^* q\boldsymbol{v}$.

Then came the study of time variation of the fields. In view of the above expression of the Lorentz force, Faraday's law must be written as $\nabla \times \boldsymbol{E} = -\gamma^*/\mu^* \partial \boldsymbol{B}/\partial t$, and Maxwell's displacement current is $\epsilon^*/4\pi \partial \boldsymbol{E}/\partial t$. The end result is that Maxwell's equations, with our notation of the three universal constants read:

$$\nabla \cdot \boldsymbol{B} = 0 \qquad \nabla \times \boldsymbol{B} = 4\pi\gamma^* \boldsymbol{J} + \gamma^*\epsilon^* \frac{\partial \boldsymbol{E}}{\partial t}$$
$$\nabla \cdot \boldsymbol{E} = \frac{4\pi}{\epsilon^*}\rho \qquad \nabla \times \boldsymbol{E} = -\frac{\gamma^*}{\mu^*}\frac{\partial \boldsymbol{B}}{\partial t} \tag{A.59}$$

In terms of the potentials, the field vectors will be $\boldsymbol{B} = \nabla \times \boldsymbol{A}$ and $\boldsymbol{E} = -\nabla V - \gamma^*/\mu^* \partial \boldsymbol{A}/\partial t$. The relation between the V and $\boldsymbol{A}$ imposed by the conservation of charge (Lorentz gauge, (A.43)) is now $\nabla \cdot \boldsymbol{A} + \gamma^*/\epsilon^* \partial V/\partial t = 0$. Most importantly, the fundamental relation (A.44) linking now *four* constants becomes

$$c^2 = \frac{\mu^*}{\gamma^{*2}\epsilon^*} \tag{A.60}$$

Why are we doing all this? Because from the set of Eqs. (A.59) and (A.60) one can easily deduce electromagnetic equation forms in all systems of units. For instance, let us retrieve Maxwell's equations for the *rationalized SI system* used in this book. The main characteristic of this system, besides using MKS mechanical units, is to define the electric charge as an *independent* unit, the Coulomb. Obviously, to obtain the set (A.48)–(A.51) from (A.59), we must set $\epsilon^* = 4\pi\epsilon$ and $\gamma^* = \mu^* = \mu_0/4\pi$. There is another, older system of units in use today, mostly by astrophysicists and laboratory plasma physicists. It is the *Gauss system*, in which the electric charge is a *derived* unit, selected in such a way that $\epsilon^* = \mu^* = 1$ without units. The unit of charge is called "statCoulomb" ($=0.33 \times 10^{-9}$ C); two charges of 1 statCoulomb at 1 cm from each other attract or repel themselves with the force of 1 dyne. In this system, the relation (A.60) demands that $\gamma^* = 1/c$. Maxwell's equations and aleatory relationships look as follows in the Gauss system:

$$\nabla \cdot \boldsymbol{B} = 0 \quad \nabla \times \boldsymbol{B} = \frac{4\pi}{c}\boldsymbol{J} + \frac{1}{c}\frac{\partial \boldsymbol{E}}{\partial t}$$
$$\nabla \cdot \boldsymbol{E} = 4\pi\rho \quad \nabla \times \boldsymbol{E} = -\frac{1}{c}\frac{\partial \boldsymbol{B}}{\partial t} \tag{A.61}$$
$$\nabla \cdot \boldsymbol{J} = -\frac{\partial \rho}{\partial t}$$
$$\boldsymbol{E} = -\nabla V - \frac{1}{c}\frac{\partial \boldsymbol{A}}{\partial t}$$
$$\boldsymbol{B} = \nabla \times \boldsymbol{A}$$

Some claim that Maxwell's equations in the Gauss system are simpler: "whenever something changes in time or moves, it has a $1/c$ in front" (in the conservation of charge, both $1/c$ factors cancel). More importantly, they point to the fact that (i) the field vectors $\boldsymbol{E}$ and $\boldsymbol{B}$ have the same units (and therefore sit together more harmoniously in the electromagnetic field tensor) and (ii) if one writes the Maxwell's equations in 4-D space-time (in which the fourth axis is $x_4 = -ct$), no universal constants at all appear. We wish to point out, however, that (i) it is most reasonable to adopt the electric charge, which obeys the perhaps most absolute conservation law of all, as an independent unit; (ii) even in relativity, space and time and the $\boldsymbol{B}$ and $\boldsymbol{E}$ components are different animals in *all* coordinate systems (never mind that they are interdependent when one system is Lorentz-transformed into another). These are some of the reasons why it is internationally recommended to adopt the SI electromagnetic unit system.

A.2 Expression for the Bounce-Average Drift Velocity

To demonstrate relation (3.3), and with it, the conservation of the second invariant J, we shall first prove (3.7) for *equipotential field lines* and static conditions. We begin by turning off the electric field and/or any external force field; the only drift will be the gradient-curvature drift (3.2). Consider Fig. A.7. When the guiding center is at point S of field line ℓ_0, the instantaneous drift displacement δx_s is directed along the binormal $\boldsymbol{b}_s$ (positive for positive particles). ℓ_Q is a field line traced through neighboring point P situated along the normal $\boldsymbol{n}_s$; ℓ_0^* is a field line traced through point S′; carefully observe the nomenclature for the infinitesimal vectors $\boldsymbol{\delta x}_s$ (drift displacement), $\boldsymbol{\delta s}_0$ (field-aligned or bounce displacement) and $\boldsymbol{\delta r}_s$ (needed for the gradient calculation). The points O, O′ and Q are field line intersections with the reference (minimum-B) surface Ω. It can be shown that in absence of field-aligned currents, field lines ℓ_Q and ℓ_0^* cut through all normals and binormals of field line ℓ, respectively, regardless of the point S,[17] including $\boldsymbol{n}_0$ and $\boldsymbol{b}_0$ on the reference surface Ω.

First we seek a general expression of the gradient vector ∇I (2.37) when it is operationally defined as described in the main text for a particle mirroring at B_m (see Fig. 3.5). This gradient will be the limit of $\delta I/\delta r_0 = (I_Q - I_0)/\delta r_0$. Obviously,

$$I_0 = \int_{s_0^*(B_m)}^{s_0(B_m)} \left[1 - B(s_0)/B_m\right]^{\frac{1}{2}} ds_0 \quad \text{and} \quad I_Q = \int_{s_Q^*(B_m)}^{s_Q(B_m)} \left[1 - B(s_Q)/B_m\right]^{\frac{1}{2}} ds_Q$$

[17] The surfaces generated by all normals or binormals of a magnetic field line are respectively tangent to two orthogonal constant Euler-potential surfaces (e.g., [5]) which intersect along ℓ_0; hence they *must* contain the field lines ℓ_Q and ℓ_0^*, respectively.

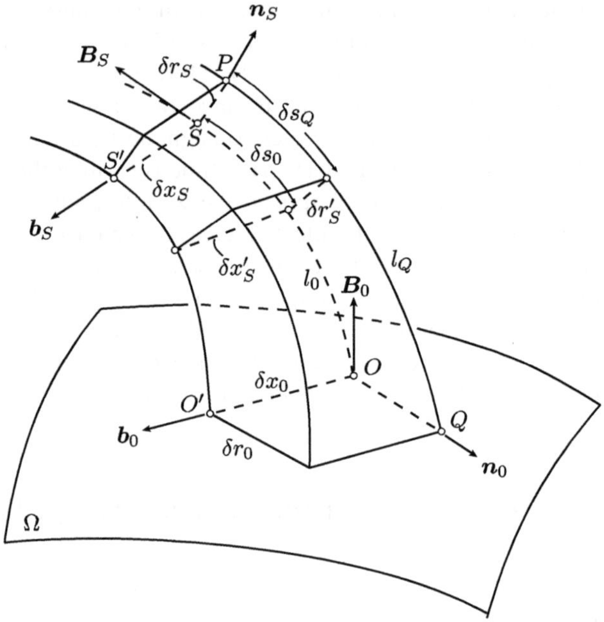

Fig. A.7 Parameters of a magnetic flux tube in the natural frame of reference

The variables in the second integral can be related to those in the first one:

$$B(s_Q) = B(s_0) - |\nabla_\perp B|\delta r_s$$

and, taking into account that in absence of currents $\nabla \times \boldsymbol{B} = 0$, $B(s_Q)ds_Q - B(s_0)ds_0 = 0$

$$ds_Q = ds_0\left(1 + \frac{|\nabla_\perp B|}{B(s_0)}\delta r_s\right)$$

Replacing these variables in the integral I_Q and expanding to first order, we obtain for the gradient:

$$|\nabla_0 I| = \frac{I_Q - I_0}{\delta r_0} = \int_{s_m^*}^{s_m} \frac{\left[1 - \frac{1}{2}B(s_0)/B_m\right]}{B(s_0)\left[1 - B(s_0)/B_m\right]^{\frac{1}{2}}} \frac{|\nabla_\perp B|\delta r_s}{\delta r_0} ds_0 \qquad \text{(A.62)}$$

In this relation $|\nabla_\perp B|\delta r_s/\delta r_0$ is dependent on the particular field geometry.

To find the expression of the drift velocity, we first note that in Fig. A.7 $B(s_0)\delta x_s \delta r_s = B_0 \delta x_0 \delta r_0$; therefore,

$$V_{0s} = \frac{\delta x_0}{\delta t} = V_s \frac{B(s_0)\delta r_s}{B_0 \delta r_0} \qquad \text{(A.63)}$$

The direction of the vector $\boldsymbol{V}_{0s}$ is that of the binormal $\boldsymbol{b}_0$, regardless of the position S of the particle. Taking this relation into account, the magnitude of the bounce-average drift velocity (3.1) is

$$\langle V_0 \rangle = \frac{2}{\tau_b} \int_{s_m^*}^{s_m} V_{0s} \frac{ds_0}{v_\parallel} = \frac{2}{\tau_b B_0 v} \int_{s_m^*}^{s_m} V_s B(s_0) \frac{\delta r_s}{\delta r_0} \frac{ds}{[1 - B(s_0)/B_m]^{\frac{1}{2}}}$$

Taking now expression (2.17) for V_s, we finally obtain

$$\langle V_0 \rangle = \frac{2p}{q\tau_b B_0} \int_{s_m^*}^{s_m} \frac{\left[1 - \frac{1}{2} B(s_0)/B_m\right]}{B(s_0)\left[1 - B(s_0)/B_m\right]^{\frac{1}{2}}} \frac{|\nabla_\perp B| \delta r_s}{\delta r_0} ds_0 \qquad \text{(A.64)}$$

The integral is identical to the above expression (A.62) for the gradient of I; hence we can write, in vector form:

$$\langle \boldsymbol{V}_0 \rangle = \frac{2p}{q\tau_b B_0^2} \nabla \boldsymbol{I}_0 \times \boldsymbol{B}_0 \qquad \text{(A.65)}$$

which is identical to the last term of (3.7).

The next task is to find an expression of the bounce-average drift velocity for a force field (ignoring for a moment the contribution from (A.65)). If $\boldsymbol{F}_s$ is the external force acting on the particle at any field line point S, taking into account (A.63) and considering that for equipotential field lines $F_s \delta r_s = F_0 \delta r_0$ (see geometry in Fig. A.7),

$$V_{0s} = \frac{F_s}{qB_0} \frac{\delta r_s}{\delta r_0} = \frac{F_0}{qB_0} \qquad \text{(A.66)}$$

This can be expressed in vector form $\boldsymbol{V}_{0s} = \boldsymbol{F}_0 \times B_0 / qB_0^2$ *independently* of the actual position S of the guiding center. The bounce-average thus turns out to be of the same form:

$$\langle \boldsymbol{V}_0 \rangle = \frac{\boldsymbol{F}_0 \times B_0}{qB_0^2}$$

Adding this expression to (A.65), we obtain the full Eq. (3.7).

Finally, we must relate this to the "official" second invariant (in the non-relativistic, equipotential field lines case),

$$J = 2pI = 2(2mM)^{1/2} \int_{s_m^*}^{s_m} [B_m - B(s_0)]^{1/2} ds_0$$

The gradient of J, which assumes constant magnetic moment M, can be decomposed into two parts: a spatial variation at constant B_m plus a variation due to the variation of B_m that stems from a change of particle energy when it drifts to a different field line:

$$\nabla_0 J = \nabla_0 J\Big|_{B_m=const} + \frac{\partial J}{\partial B_m}\nabla_0 B_m$$

The first term is, by definition, equal to $2p\nabla_0 I$. If U is the potential of the force F, taking into account that $M = (W - U)/B_m$ the gradient of B_m will be

$$\nabla_0 B_m|_{M=const} = -\frac{\nabla_0 U}{M} = \frac{F_0}{M}$$

Note that $\nabla_0 B_m$ is *not* the gradient of the local field and that in absence of external forces $\nabla_0 B_m = 0$.

On the other hand, from the above expression of J and (2.35),

$$\frac{\partial J}{\partial B_m} = (2mM)^{1/2}\int_{s_m^*}^{s_m}\frac{ds}{[B_m - B(s_0)]^{1/2}} = \left(\frac{2mM}{B_m}\right)^{1/2} S_b$$

The term arising from the derivation of the integral limits is zero. Combining these results we obtain

$$\nabla_0 J = 2p\nabla_0 I + \tau_b \boldsymbol{F}_0 \tag{A.67}$$

It is important to note that Eq. (3.3) holds in the most general case, even if field lines are not equipotentials and fields and forces vary in time under adiabatic conditions (2.1) and (2.2).

As a *first corollary*, an important note on the above relation (A.66). If the external force at a given point S of the magnetic field line is an electric field $\boldsymbol{E}$ (and as expected, the field lines are equipotentials), the projected drift velocity of the reference point O $\boldsymbol{V}_{0s} = 1/B_0\boldsymbol{E}_0 \times \boldsymbol{e}_0$ is the *same for all points* S. This means that if we place 90° pitch angle particles of near-zero energy all along a given field line, after a time δt they will all still be on a common guiding field line. This is crucial for the definition of field line velocity ((1.38) and page 20).

As the *second corollary*, with a bit of boring vector algebra using the above equations, one can show that all particles of given magnetic moment and J or I value whose guiding centers are contained at one given time in one given field line tube of flux $\Delta\Phi$ (like the one shown in Fig. A.7) will always occupy flux tubes of the *same flux value* during their drift. In other words, not only is the flux through a cyclotron orbit conserved, also the magnetic flux of a bundle of guiding field lines of particles of the same M and J remains constant.

A.3 Conservation of the Third Adiabatic Invariant

We shall prove the conservation theorem of the third adiabatic invariant (3.38) *for equatorial non-relativistic particles only*. Consider an initial guiding drift shell Σ in a stably-trapping magnetic field. For equatorial particles any guiding drift shell is reduced to a closed $B = \text{const.}$ curve on the minimum-B surface, as shown in Fig. A.8. Remember that for a dipole-like geometry (which has a singularity at the origin) the flux to be considered in (3.38) is that *outside* the drift contour[18] (Fig. 3.22a). In general, the drift period is given by $\tau_d = \oint_\Sigma d\ell / V_D$ (3.35) with a drift speed which in our case is $V_D = (T/qB^2)\nabla_\perp B$ (relation (2.15); in the figure we have assumed positive particles and an inward-pointing field gradient). Let us assume that at time t_P the magnetic field begins to increase adiabatically, such that condition (3.36) is fulfilled. At time t_Q, $\boldsymbol{V}_D$ and the electric drift $\boldsymbol{V}_E = (\boldsymbol{E}_{ind} \times \boldsymbol{B})/B^2$ will have brought the particle's guiding center to point Q, as shown in Fig. A.9, which is very close to point Q* because of adiabatic condition (3.36) (a negative particle would gradient-drift in the opposite direction, but the induced electric field drift would still be outwards). The differential vector $\boldsymbol{\Delta r}$ is perpendicular to Σ. The contour Σ_Q shown in the figure is the guiding drift shell at that instant, i.e., the drift path if we were to stop the time variation at time t_Q. The change in B would be:

$$\Delta B = B(\boldsymbol{r} + \Delta\boldsymbol{r}, t + \Delta t_{PQ}) - B(\boldsymbol{r}, t) = (\partial B/\partial t)\Delta t_{PQ} - \nabla_\perp B \Delta r \qquad \text{(A.68)}$$

where $\Delta t_{PQ} = t_Q - t_P$; the minus sign in (A.68) reflects the field geometry and its time variation adopted for Fig. A.8. Note that this is a purely *local* relationship; it can be written for *any* point of the initial contour Σ.

Now we calculate the *change* in the magnetic flux. If Φ is the magnetic flux *outside* the initial drift contour Σ, the induced electric field during the time-variation will be related to the changing magnetic field through Faraday's law:

$$\frac{\partial \Phi}{\partial t} = -\oint_\Sigma \boldsymbol{E}_{ind} \cdot \boldsymbol{d\ell}$$

with the sense of $\boldsymbol{d\ell}$ prescribed by the right-hand rule (as applied to the portion outside of the contour), and Σ considered as fixed in space—which it is anyway by definition of guiding drift shell. Since we have assumed that the magnetic field is increasing ($\dot{B} > 0$) (Fig. A.8), we'll have $\oint \boldsymbol{E}_{ind} \cdot \boldsymbol{dl} < 0$, i.e., $\boldsymbol{E}_{ind}$ indeed points counterclockwise as shown, causing an additional *outward* drift—precisely, the drift

[18] In the field geometry of a mirror machine (see Fig. 2.5) the magnetic flux to be considered is that *inside* the more or less cylindrical drift shells. This would reverse the sign of the final result (A.74) in this appendix.

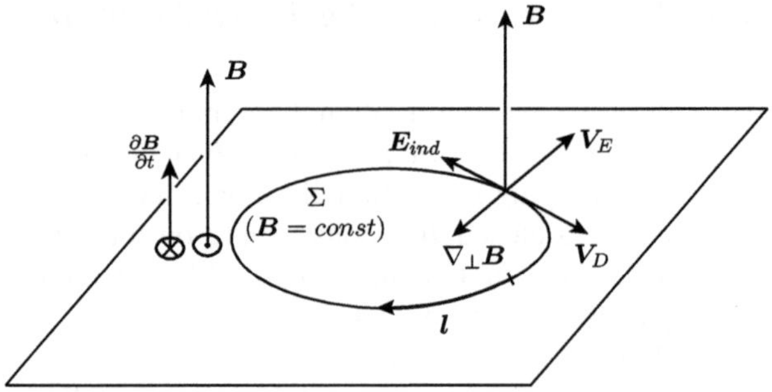

Fig. A.8 Sketch of relative directions of key vectors for a positive particle drifting in a slowly increasing magnetic field

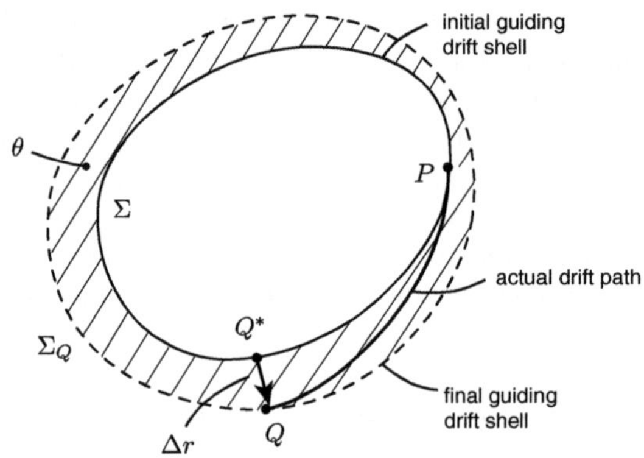

Fig. A.9 Actual drift path of a particle during a time interval $\Delta t_{PQ} < \tau_d$

responsible for the path of the guiding center from P to Q. More specifically, we now compute the change $\Delta\Phi_{PQ}$ of the magnetic flux once the particle has arrived at point Q. By definition, it is the flux delimited by the new contour Σ_Q in the *final* field minus the flux delimited by the initial contour Σ in the *initial* field: $\Delta\Phi_{PQ} = \Phi_{\Sigma_Q}(t + \Delta t) - \Phi_\Sigma(t)$. According to the figure, to first order this is equal to the time-change of the flux $\Delta\Phi_\Sigma$ for the *fixed* contour Σ in the time-dependent field, *minus* the flux through the strip Θ between the two contours for a time-constant field. Therefore,

$$\Delta\Phi_{PQ} = \Delta\Phi_\Sigma - \Delta\Phi_\Theta \tag{A.69}$$

where

$$\Delta\Phi_\Sigma = -\Delta t_{PQ} \oint_\Sigma \boldsymbol{E}_{ind} \cdot \boldsymbol{\delta\ell} = \Delta t_{PQ} \oint_\Sigma E_{ind} \delta\ell$$

$$\Delta\Phi_\Theta = B \oint |\boldsymbol{\Delta r} \times \boldsymbol{\delta\ell}| = B \oint \Delta r \delta\ell \tag{A.70}$$

where the magnitude of the segments $\boldsymbol{\Delta r}$ ($\perp$ $\boldsymbol{\delta\ell}$) can be obtained from relation (A.68) above[19]:

$$\Delta r = \frac{1}{\nabla_\perp B}\left(\frac{\partial B}{\partial t}\Delta t_{PQ} - \Delta B\right) = \frac{T}{qB^2}\left(\frac{\partial B}{\partial t}\frac{\Delta t_{PQ}}{V_D} - \frac{\Delta B}{V_D}\right) \tag{A.71}$$

To integrate this equation along the contour Σ as prescribed in (A.70), we must take into account the fact that in (A.71), only $\nabla_\perp B$ (i.e., V_D) and $\partial B/\partial t$ are functions of ℓ. We obtain

$$\Delta\Phi_\Theta = \frac{T}{qB}\left(\Delta t_{PQ}\oint_\Sigma\left(\frac{\partial B}{\partial t}\right)\frac{d\ell}{V_D} - \tau_d\Delta B\right) \tag{A.72}$$

What we now need is to find an expression for ΔB, the change of B along the *actual* transient drift path PQ of the particle that ended up on Σ_Q when the time variation was stopped at t_Q. For that purpose, we relate ΔB to the energy change ΔT_{PQ} experienced by the particle, using the conservation of the first invariant: $\Delta B = M\Delta T_{PQ}$. The result is

$$\Delta\Phi_\Theta = \frac{M}{q}\Delta t_{PQ}\oint\left(\frac{\partial B}{\partial t}\right)\frac{d\ell}{V_D} - \frac{\tau_d}{q}\Delta T_{PQ} \tag{A.73}$$

Inserting $\Delta\Phi_\Sigma$ and $\Delta\Phi_\Theta$ in (A.69) and dividing by Δt_{PQ}, we finally have for the differential quotient:

$$\left.\frac{\Delta\Phi}{\Delta t}\right|_{PQ} = \int_{t_P}^{t_P+\tau_d} EV_D dt + \frac{\tau_d}{q}\frac{\Delta T_{PQ}}{\Delta t_{PQ}} - \frac{1}{q}\int_{t_P}^{t_P+\tau_d} M\frac{\partial B}{\partial t}dt$$

In this relation, note that ΔT_{PQ} is the integral over time of the time-rate of change of the particle's energy between t_P and t_Q, so that

$$\left.\frac{\Delta\Phi}{\Delta t}\right|_{PQ} = \frac{\tau_d}{q}\int_{t_P}^{t_Q}\frac{dT}{dt}dt - \frac{1}{q}\int_{t_P}^{t_P+\tau_d}\left[M\frac{\partial B}{\partial t} - qEV_D\right]dt$$

[19]Here we come to one of those typical intricacies of adiabatic theory. In the above equations we have two kinds of differentials: *mathematical* or *true* differentials, which are supposed to tend to zero in order to validate the operation in which they participate (a derivative, an integral, a differential quotient, or just an algebraic relationship). In this book we usually denote them with a "d" or "∂", like in the expression of Faraday's law above. And then we have *physical* differentials, which represent very small quantities (compared to typical values of intervening variables) but still must remain finite. We usually indicate them with "δ" or "Δ". An example of the latter is $\boldsymbol{\Delta r}$ in (A.70), which for each point on the drift contour is "physically" a vector sum (integral) of higher order differential vectors $(\boldsymbol{E}_{ind} \times \boldsymbol{B}/B^2)\delta t$, all perpendicular to the guiding drift contour Σ (see Fig. A.8). But even the higher order differentials like this δt have lower limits, in this case imposed by the condition that $\delta t \gg \tau_C$!

Considering the general energy equation (2.26), we realize that the second integral is just the time-integral of dT/dt over a *full* drift cycle! In summary, the final expression of the rate of change of Φ can be written:

$$\left.\frac{\Delta\Phi}{\Delta t}\right|_{PQ} = \frac{\tau_d}{q}\left[\frac{1}{(t_Q - t_P)}\int_{t_P}^{t_Q}\frac{dT}{dt}dt - \frac{1}{\tau_d}\int_{t_P}^{t_P+\tau_d}\frac{dT}{dt}dt\right]$$

The two quantities between the brackets are the *drift-average values*[20] of the particle's kinetic energy rate of change—in other words the power delivered (or taken) by the induced electric field in the cyclotron and drift motions, during the partial drift between points P and Q and one full drift turn, respectively:

$$\left.\frac{\Delta\Phi}{\Delta t}\right|_{PQ} = \frac{\tau_d}{q}\left[\left\langle\frac{dTX}{dt}\right\rangle_{PQ} - \left\langle\frac{dT}{dt}\right\rangle_{\Sigma}\right] \tag{A.74}$$

This is an interesting result: we may even call it a temporal type of shell splitting! Whenever $\Delta t_{PQ} = n\tau_d$, which means that the particle has drifted exactly an integer multiple of times around, the change of the drift shell flux value is *exactly zero*, any dependence on the initial contour point P disappears and *all* particles on the original drift contour Σ will assemble again on one and the same shell. This is no longer true at any other times $t_Q \neq n\tau_d$, at which the drift contour appears split into a fuzzy continuum, until another integer multiple of τ_d is reached. We may summarize this in the relation

$$\lim_{\Delta t \to n\tau_d}\frac{\Delta\Phi}{\Delta t} \equiv 0 \tag{A.75}$$

provided that $n\tau_d \ll \Delta B/(dB/dt)$. This represents a "pulsating sharpness" of drift shells in a slowly changing magnetic field. Of course, because of the adiabatic condition (3.36) all these drift shells will be very close to each other; still this can lead to radial diffusion under certain circumstances [6]. It also allows us to declare that *on the average* over many drift times, the third invariant is conserved to first adiabatic order. For azimuthally symmetric $\partial B/\partial t$ and $\boldsymbol{V}_D$, the two averages

[20] Throughout this text we use "phase averages" of some quantity X. Let us summarize them in one place in orderly fashion:

Cyclotron average: $\langle X\rangle_c = (1/\tau_c)\oint X d\phi/\dot{\phi}$; $\tau_c = \oint d\phi/\dot{\phi}$
Bounce average: $\langle X\rangle_b = (1/\tau_b)\oint X ds/\langle v_\parallel\rangle_c$; $\tau_b = \oint ds/\langle v_\parallel\rangle_c$
Drift average: $\langle X\rangle_d = (1/\tau_d)\oint X dl/\langle V_D\rangle_b$; $\tau_d = \oint dl/\langle V_D\rangle_b$

For variable l see Fig. A.8; the integrals are to be taken along closed geometric entities: respectively, the Larmor circle in the GCS (*not* the actual cyclotron orbit), the guiding field line (*not* the actual field lines occupied by the bouncing particle), and the guiding drift shell (*not* the actual drift shell engendered by the drifting particle in a changing field). Note that the bounce average contains a cyclotron average, and that the drift average contains the bounce average of a cyclotron average!

in (A.74) are identical, (A.75) is satisfied at all times and particles stay on the same shell at all times during an adiabatic time variation. Although we have limited ourselves to equatorial particles (for which noon-midnight, dawn-dusk asymmetry effects would be maximum anyway), it can be shown that the above conclusions hold for arbitrary pitch angles [7, 8].

References

1. C. Cercignani, *Vorlesungen zu Maxwells Theorie der Elektrizität und des Lichts*, Bd. II (Oxford University Press, Oxford, 2010)
2. J.A. Stratton, *Electromagnetic Theory* (McGraw-Hill, New York/London, 1941)
3. L. Boltzmann, *Ludwig Boltzmann, The Man who Trusted Atoms* (Metzger und Wittig, Leipzig, 1893)
4. V.M. Vasyliūnas, Relation between magnetic fields and electric currents in plasmas. Ann. Geophys. **23**, 2589–2597 (2005)
5. D.P. Stern, Euler potentials. Am. J. Phys. **38**, 494–501 (1970)
6. S. Lejosne, Modélisation du phénomène de diffusion radiale au sein des ceintures de radiation terrestres par technique de changement d'échelle. PhD thesis, Université de Toulouse, France, 2013
7. T.G. Northrop, E. Teller, Stability of the adiabatic motion of charged particles in the Earth's field. Phys. Rev. **117**, 215–225 (1960)
8. T.G. Northrop, *The Adiabatic Motion of Charged Particles* (Interscience Publishers, New York, 1963)

Index

J.G. Roederer and H. Zhang, *Dynamics of Magnetically Trapped Particles*, Astrophysics and Space Science Library 403, DOI 10.1007/978-3-642-41530-2,

MIX
Papier aus verantwortungsvollen Quellen
Paper from responsible sources
FSC® C105338

FSC
www.fsc.org

If you have any concerns about our products,
you can contact us on
ProductSafety@springernature.com

In case Publisher is established outside the EU,
the EU authorized representative is:
Springer Nature Customer Service Center GmbH
Europaplatz 3, 69115 Heidelberg, Germany

Printed by Libri Plureos GmbH
in Hamburg, Germany